SpringerBriefs in Physics

More information about this series at:
http://www.springer.com/series/8902

V.A.G. Rivera • O.B. Silva • Y. Ledemi
Y. Messaddeq • E. Marega Jr.

Collective Plasmon-Modes in Gain Media

Quantum Emitters and Plasmonic Nanostructures

 Springer

V.A.G. Rivera
Institute of Physics of São Carlos - USP
São Carlos, São Paulo
Brazil

O.B. Silva
Institute of Physics of São Carlos - USP
São Carlos, São Paulo
Brazil

Y. Ledemi
Center for Optics, Photonics and Laser
Laval University
Québec, QC
Canada

Y. Messaddeq
Center for Optics, Photonics and Laser
Laval University
Québec, QC
Canada

E. Marega Jr.
Institute of Physics of São Carlos - USP
São Carlos, São Paulo
Brazil

ISSN 2191-5423   ISSN 2191-5431 (electronic)
ISBN 978-3-319-09524-0   ISBN 978-3-319-09525-7 (eBook)
DOI 10.1007/978-3-319-09525-7
Springer Cham Heidelberg New York Dordrecht London

Library of Congress Control Number: 2014945744

Printed on acid-free paper

Springer is part of Springer Science+Business Media (www.springer.com)

*To Nina, Sami, and Maritza
from V.A.G. Rivera*

Preface

When we consider for a few minutes the new scientific and technological advances available today, there is no doubt that we are witnessing unparalleled scientific and technological progress, greater than any made by humanity in all its history, and that this progress is thanks to continuing scientific research. It is true, however, that we still have much to understand and explain, and we will only succeed if we continue on the present path of scientific development toward the improved welfare of society. Particularly exciting in this regard is the continuing miniaturization of electronic devices such as transistors; these electronic components can be found in all electronic devices and are currently manufactured at sizes on the order of microns to nanometers.

When we decided to write this book, we wanted to reflect on the importance of plasmonics in optical gain materials, especially on the use of rare-earth ions widely used in photonics. This concise volume does not endeavor to provide a comprehensive review of devices or photonic materials, but rather an overview of potential applications resulting from combining the two major areas of photonics and plasmonics, leading to a new and exciting area of scientific research—nanophotonics.

The main objective of this book is to give a more accurate picture of quantum-plasmonic interaction in materials with high optical gain (rare-earth-doped materials) and metallic nanostructures that allow us direct manipulation of the energy levels of the rare-earth ions. These allow us to explore the interaction of light and matter at the nanoscale and its application in the emerging field of nanophotonics.

This book is divided into four main chapters. Chapter 1 discusses in detail the interaction of light with metallic nanostructures, the formation and propagation of a surface plasmon, and how these plasmons interact in a gain medium. These topics are analyzed from an electrodynamics perspective. Additionally, we present the optical properties of rare-earth ions from a quantum point of view, as well as some of the main applications of these properties in the field of photonics. In Chap. 2, we conduct a more detailed study of metallic nanoparticles, their main properties, and the characteristics of their types, as well as the interaction effects of localized surface plasmon resonance on rare-earth ions and other quantum emitters. Chapter 3 presents a study of the interaction of plasmonic nanostructure arrays with rare-earth ions and other quantum emitters. In both chapters, we present several experimental verifications and physical interpretations of the theories compiled here. Moreover, we include both quantum and electrodynamics treatments of these interaction processes since, after all, we want to provide a basis for these equations and interpretations that readers can use to explore them in greater detail. Finally, Chap. 4 presents a clear and concise vision of the future applications and challenges of nanophotonics, based on plasmon–photon interactions and the rapid development of photonics, plasmonics, and nanophotonics.

This book is the product of many months of collaborative work, which was, in general, a very pleasant experience. It was written by recognized experts in their areas of expertise, from two excellent research centers, where there is a well-founded tradition of international research, the Instituto de Física de São Carlos, Universidade de São Paulo, Brazil, and the Centre d'optique, photonique et laser, Université Laval, Canada.

We hope that this text will serve its purpose and provide a useful tool for current and future participants in the new field of nanophotonics. We also hope to strengthen the sense of community among the different scientific subfields discussed in this book.

São Carlos, São Paulo, Brazil   V.A.G. Rivera
São Carlos, São Paulo, Brazil   O.B. Silva
Québec, QC, Canada   Y. Ledemi
Québec, QC, Canada   Y. Messaddeq
São Carlos, São Paulo, Brazil   E. Marega Jr.

Acknowledgments

This work was supported by the Brazilian agencies Coordenação de Aperfeiçoamento de Pessoal de Nível Superior (CAPES), Fundação de Amparo à Pequisa do Estado de São Paulo (FAPESP) (Processes: 2009/08978-4 and 2011/21293-0), and the Conselho Nacional de Desenvolvimento Científico e Tecnológico (CNPq). The Instituto Nacional de Óptica e Fotônica (INOF) (Processes: CNPq 573587/2008-6, and FAPESP 2008/57858-9) and the Centro de Pesquisa em Óptica e Fotônica (CEPOF) (Process: FAPESP 2013/07276-1) are also acknowledged, as well as the Núcleo de Apoio a Pesquisa em Óptica e Fotônica (NAPOF) at the Universidade de São Paulo.

This work also was supported by the Canadian agencies of the Canadian Excellence Research Chair program (CERC) on Enabling Photonic Innovations for Information and Communication, the Natural Sciences and Engineering Research Council of Canada (NSERC), and the Canada Foundation for Innovation (CFI), as well as the Quebec agency Fonds Québecois de la Recherche sur la Nature et les Technologies (FQRNT). They are acknowledged for their financial support.

The authors are also grateful for the work facilities provided by the Instituto de Física de São Carlos of Universidade de São Paulo at São Paulo City, Brazil, and by the Center for Optics, Photonics and Lasers at Laval University in Quebec City, Canada.

The authors are also grateful to their respective families for their patience and unswerving support throughout the writing process for this book.

In particular, V.A.G. Rivera is deeply grateful to all colleagues, experts in various areas of plasmonics and photonics, for their effort and their willingness to share their recent knowledge within the framework of this book. With this, we hope that this book will serve its purpose and provide a useful tool for future researchers in this area and the different subfields.

Contents

List of Figures

List of Symbols

A^u	Quadrivector potential
a and b	Electric field components for the forward and backward electromagnetic component, respectively
a_n and b_n	Coefficients from Mie theory
a_w and $a_w^\dagger$	Creation and annihilation operators
$\beta = \frac{\pi d n_d}{\lambda}$	Size parameter of the nanoparticle
$B_{k,q}$	Crystal field parameter
$\boldsymbol{B}$	Magnetic induction
B	Modal attenuation
γ	Damping factor
γ_{ji}	Damping factor related to transition $i \rightarrow j$
γ_{exc}	Excitation rate $(=\lvert p \cdot \boldsymbol{E} \rvert^2)$
γ_{em}	Emission rate
Γ_{rad}	Emission, radiative
Γ_{nonrad}	Emission, nonradiative
Γ_{LSPR}	Energy transfer rate to the LSPR mode
c	Speed of light in vacuum
$C_{k,q}$	Tensor operator
$\delta(\boldsymbol{r} - \boldsymbol{r}')$	Dirac delta function, spatial
$\delta(t' - t)$	Dirac delta function, temporal
$\delta, \delta(\omega)$	Skin depth
$\boldsymbol{D}$	Dielectric displacement
d	Size of the nanoparticle
d'	Hole size
ε_0	Vacuum permittivity
$\varepsilon(\omega), \varepsilon_{\text{m}}(\omega)$	Metal electric permittivity
ε_{d}	Dielectric permittivity
ε_{R}	Relative permittivity of a material
$\text{Im}[\varepsilon_{\text{m}}(\omega)]$	Imaginary part of dielectric metal

$\mathrm{Re}[\varepsilon_m(\omega)]$	Real part of dielectric metal
e	Electron charge
$\mathbf{E}$	External electric field or applied electric field
$\mathbf{E}_{\mathrm{QE}}$	Electric field from the quantum emitter
$\mathbf{E}_{\mathrm{m}}$	Electric field on metal
$\mathbf{E}_{\mathrm{loc}}$	Local electric field
$\mathbf{E}_{\mathrm{eff}}$	Effective electric field
$\mathbf{E}_{\mathrm{d}}$	Electric field on dielectric
$\mathbf{E}_0$	Amplitude of the electric field of incident light
$\epsilon_A(\lambda)$	Extinction coefficient function of the acceptor in the λ to $\lambda+\Delta\lambda$ region
$\dfrac{F_{\mathrm{p}}}{\tau_{\mathrm{r}}}$	Purcell-enhanced spontaneous emission rate
$F_D(\lambda)$	Corrected luminescence intensity function of the donor in the λ to $\lambda+\Delta\lambda$ region
f_{ji}	Oscillator strength
g	Electromagnetic density of states
$G(r, r', t' - t)$	Green's function
η	Quantum yield of the luminescence
η_D	Energy transfer efficiency
η_{LSPR}	Coupling efficiency between rare-earth ion and nanoparticle
η_0	Internal quantum efficiency of the energy transfer between rare-earth ion and nanoparticle
h	Planck constant
$\hbar$	Reduced Planck constant
$\mathbf{H}$	Magnetic field
H	Hamiltonian operator
H_{QE}	Magnetic field from the quantum emitter
$\mathrm{H}_{n + 0.5}(x)$	Half-integer-order Hankel function of the second kind
H_{DC}	Hamiltonian of the dynamic coupling
θ	Polar angle
$\Theta_{w,\mathbf{k},\alpha}$	Possible projection of the vectorial amplitude of the plasmon
I_D	Intensity of emission of a rare-earth ion or nanoparticle (donor)
$I_{\mathrm{NP}}(\omega)$	Intensity of radiation from the nanoparticle
$I_{D + A}$	Intensity of emission of a rare-earth ion and a nanoparticle
i	Imaginary number
$\mathbf{j}, \mathbf{J}_{ext}$	Current density and external current density
J	Total angular momentum
J_m	Bessel function of order m
$J_{n+1/2}(z)$	Half-integer-order Bessel function
$J(\lambda)$	Degree of spectral overlap between the donor emission and the acceptor absorption
$\kappa = l/2r$	Aspect ratio of the nanoparticle
	Wave vector for free space

$k_0 \left(= \frac{\omega}{c} = \frac{2\pi}{\lambda} \right)$

k_{spp}	Wave vector for the surface plasmon polariton
$\boldsymbol{k}$	Wave vector
k	Constant associated with a restoring force
k_{r}	Radiative decay rate
k_{nr}	Nonradiative decay rate
λ_{exc}	Wavelength of the light beam excitation
$\lambda_{\mathrm{Wood}}(i,j)$	Wavelength of the Wood's anomaly
λ_{SPP}	Wavelength of surface plasmon polariton waves
L_{SPP}	Distance that the surface plasmon polariton waves propagate along the interface
L_{i}	Geometrical factor
l	Electron mean free path
μ_0	Vacuum permeability
u_α	Polarization of the photon
μ_d	Relative permeability
$\vec{u}_{xy}$	Unit vector in the plane of the film in the direction of the projection of the incident light wave vector
m, m^*	Effective electron mass
m_{NP}	Nanoparticle mass
ν	Frequency
N	Avogadro's number
N	Number of electrons in the f electron shell of a rare-earth ion
N_i and N_j	Electronic populations of the initial and final states
$\boldsymbol{n}$	Unit vector in the direction of the point "p" of interest
$\hat{n}$	Normal unitary vector under the interface
n_j	Density of the conduction electrons analogous to $s(\boldsymbol{r},t)$
n_{eff}	Effective refractive index
n_{d}	Refractive index of the dielectric
$\tilde{n} = n(\omega) + i\kappa(\omega)$	Refractive index
o^2	Relative orientation in space of the transition dipoles of the donor and acceptor
$\boldsymbol{P} = np$	Macroscopic polarization, where n is the number of carrier charges per volume
P_{strength}	Oscillator strength
$P(A)$	Probability of finding an acceptor ion at a distance from the donor between r and $r+\Delta r$
$P_l(\cos\theta)$	Legendre polynomials
$\boldsymbol{p}$	Electric dipole momentum
$p_{\mathrm{NP}j}$	Electric dipole moment of the nanoparticle
p_{QE}	Dipole moment of the quantum emitter
$q_{\mathrm{w}}(\boldsymbol{r})$	Amplitudes of oscillation charges, eigenmodes
Q	Quality factor of the cavity
q	Emission probability

γ	Quantum yield of the donator
$\rho = \sqrt{x^2 + y^2}$	Distance vector, where x, y, and z are the Cartesian coordinates
ρ, ρ_{ext}	Charge density and external charge density
$\rho_e(\omega)$	Local density of optical states of the quantum emitter
$\rho_c(\omega)$	Local density of optical states of the electromagnetic environment
$\rho(R)$	Crystal field charge density
R	Radius of the nanoparticle
$\boldsymbol{r}$	Position vector
$\boldsymbol{r}_0$	Amplitude of the position vector
$\boldsymbol{r}_{ij}$	Separation distance between nanostructure and rare-earth ion
$'r$	Reflection coefficient
σ_{ext}	Extinction cross section of a spherical nanoparticle
$\sigma(\omega)$	Absorption cross section
s	Displacement of the volume charge
$\dot{s}$	Associated velocity of the volume charge
S	Refractive index sensitivity
S_{ED}	Electric dipole intensity
S_{MD}	Magnetic dipole intensity
τ	Characteristic time of the free electron
$\tau_{uncoupled}$	Lifetime of uncoupled quantum emitter
τ_r, τ_{nr}	Radiative and nonradiative lifetime decay lifetimes of the quantum emitter
$\tau_{coupled}$	Lifetime of the quantum emitter coupled with a metallic nanostructure
τ_D	Radiative lifetime of the donor ion
T	Kinetic energy
T_r	Transmission coefficient
$T_{w,\mathbf{k},\alpha}$	Probability rate from the nanoparticle
t	Time
Φ	Phase
$\phi(r, \theta)$	Scalar potential
U	Potential energy
V	Volume of the nanoparticle
V_{eff}	Mode volume
V_{EF}	Interaction of the ion with the electromagnetic field
V'	Interaction volume
v_F	Fermi velocity
W_{DA}	Rate of the transfer process
ω	Frequency of the incident light
ω_{spp}	Frequency of the surface plasmon polariton
ω_{qe}	Transition frequency of the quantum emitter
ω'_p	Plasma frequency

ω_p	Plasma frequency for free electron gas
ω_{ji}	Frequency of transitions of electrons between the energy levels E_j and E_i
$\omega_0 = \sqrt{\frac{k}{m}}$	Natural frequency of oscillation of the bound electron
χ	Factor that depends on the refractive index of the medium in which the REIs are embedded
$\chi(\omega)$	Electric susceptibility
$\zeta(r_i)$	Spin–orbit coupling function
Y_{kq}	Spherical harmonics
Ψ and ξ	Riccati–Bessel functions
Ψ_w and Π_w	Functions of creation and destruction operators
$\Box = \nabla^2 - \dfrac{1}{c^2}\dfrac{\partial^2}{\partial t^2}$	D'Alembertian operator

Abbreviations

CMOS	Complementary metal oxide semiconductor
CQED	Cavity quantum electrodynamics
DDA	Discrete dipole approximation
ED	Electric dipole
EM	Electromagnetic
EOT	Extraordinary optical transmission
FDTD	Finite-difference time domain
FEM	Finite element method
ISP	Induced surface plasmon
LDOS	Local density of optical states
LSP	Localized surface plasmon
LSPR	Localized surface plasmon resonance
MD	Magnetic dipole
MIR	Middle infrared
NIR	Near infrared
NP	Nanoparticle
QD	Quantum dot
QE	Quantum emitter
REI	Rare-earth ion
SERS	Surface-enhanced Raman scattering
SNOM	Scanning near-field optical microscopy
SP	Surface plasmon
SPASER	Surface plasmon amplification by stimulated emission of radiation
SPP	Surface plasmon polariton
SPR	Surface plasmon resonance
STM	Scanning tunneling microscopy
TE	Transverse electric
TM	Transverse magnetic
UV	Ultraviolet

Chapter 1
Quantum Aspects of Light–Matter Interaction

1.1 Introduction

The interactions of light with matter have been heavily investigated and well described from the end of the nineteenth century and throughout the twentieth century. By now, the phenomena of light propagation in matter, such as absorption, nonlinear absorption, refraction, reflection, dispersion, etc., are considered to be perfectly well understood and described from macro- and micro-scale points of view. For a comprehensive description of the phenomenology of light–matter interaction, the reader is invited to read the major references [1, 2]. The late twentieth century has witnessed major breakthroughs and significant progress in the fields of fabrication, observation, and characterization of objects at incredibly small sizes, giving rise to a whole set of new terms with the prefix "nano-" [3–7]. This rapid emergence has resulted in many major advances in scientific fields related to nano-objects (like nanoparticles, nanorods, nanofilms) or nanostructured materials.

In addition to the great scientific interest devoted to semiconductors, nanocrystals and quantum dots [8], the interaction of light with metal nano-objects associated with a dielectric media has also attracted a great deal of attention since the beginning of the twenty-first century, owing to its extraordinary features. Indeed, when some metals, particularly the noble metals (Ag, Au, Pt) but also base metals (Cu, Al, etc.) [9] are embedded or surface-deposited at a nanoscale in a dielectric medium, they exhibit a unique ability under certain conditions to generate electromagnetic waves, so-called surface plasmons, propagating along the metal–dielectric interface. These surface plasmons possess a unique ability to generate a local electromagnetic field around the nanostructure. This phenomenon can then be exploited to enhance the spectroscopic features of rare-earth ions, such as luminescence; some of their energy levels are resonant with those of plasmons [10]. The history and extraordinary features of plasmons, as well as their various fields of application, will be described later.

© V.A.G. Rivera, O.B. Silva, Y. Ledemi, Y. Messaddeq, and E. Marega Jr. 2015
V.A.G. Rivera et al., *Collective Plasmon-Modes in Gain Media*, SpringerBriefs
in Physics, DOI 10.1007/978-3-319-09525-7_1

The quantum theory of light is a useful tool for describing microscopic interactions between light and matter. The electromagnetic state is represented by photon number states while the electromagnetic field becomes an operator. Such a description of light provides a quantitative description of absorption, spontaneous and stimulated emission of photons by a two-level system. In particular, it allows the derivation of a quantitative treatment of light amplification. It also predicts pure quantum effects, such as photon coalescence or antibunching. The quantum theory of light can be extended to nondispersive and nonlossy media. Each photon in the material corresponds to the excitation of a mode characterized by a wave vector k and a circular frequency ω, such as $k = \frac{n.\omega}{c}$, where n is the refractive index of the medium and c is the speed of light in a vacuum. It is the purpose of this paper to introduce a quantification scheme for surface waves propagating along an interface.

Over the past 20 years, there has been greatly increasing interest in fundamental research on the one hand and on potential applications in terms of implementation and integration in devices on the other hand. Surface plasmons are, by definition, coherent oscillations of charges propagating along a metal–dielectric interface and they behave almost like free-electron plasma [10]. Well before the interaction of light with metal nanostructures became a topic of interest for scientists, their optical peculiarities were exploited to produce bright colorations in ancient glasses and stained glasses, for instance. A well-known example illustrating this interaction of light with a metal–dielectric interface is the Lycurgus cup (Byzantine Empire, fourth century BC, displayed at the British Museum) which presents a bright red color when illuminated in transmission and appears green when illuminated in reflection. This peculiar behavior is due to the presence in the glass matrix of trapped gold nanoparticles, which strongly absorb light in the green part of the visible spectral range.

1.1.1 Historical Survey

The first scientific observations of surface plasmons were reported at the beginning of the twentieth century. In 1902, Professor R.W. Wood observed an unexplained behavior in optical reflection measurements of metallic gratings [11]. At the same time, in 1904, Maxwell Garnett described the bright colors observed in metal-doped glasses by using the Drude theory of metals and the electromagnetic properties of small spheres studied by Lord Rayleigh [12]. While trying to better understand this phenomenon, Gustav Mie developed, in 1908, his theory on light scattering by small spherical particles [13]. This theory is still commonly employed nowadays. Fifty years later, in 1956, David Pines described theoretically the characteristic energy losses of rapid electrons propagating across metals [14]. He attributed these losses to collective oscillations of free electrons on the metal. By analogy with the earlier works achieved on plasma oscillations in electrical discharges in gas, he named these oscillations "plasmons." The same year, Robert Fano introduced the term "polariton" to describe coupled oscillations of bonding electrons with light inside a transparent material [15]. In 1968, about 60 years after the first observations made by Robert Wood, R. H. Ritchie and his colleagues reported the abnormal behavior of metal gratings in terms of excited surface plasmon resonance modes [16]. A major breakthrough is surface plasmon studies occurred the same year when Andreas Otto, Erich Kretschmann, and Heinz Raether presented methods to optically excite surface plasmons on metal surfaces [17, 18]. These pioneering works then opened the way for scientists to exploit and take advantage of this phenomenon in their work. Since then, studies in this field have yielded clear evidence for the coupling between electron oscillations and the electromagnetic field. The term "Surface Plasmon Polariton" (SPP) was first introduced by Stephen Cunningham et al. in 1974 [19].

Since the emergence of the new domain of surface plasmon optics, we have observed a progressive transition from fundamental studies toward applied research. This new research trend is growing at the same time that technologies such as optical lithography, optical data storage, and even high-density electronic devices are approaching the limits of fundamental physics. Several technological barriers can be overcome through the utilization of the exclusive properties of surface plasmons. Thanks to numerous recent studies, a wide domain of knowledge and techniques based on plasmon optics is currently under development, including a large variety of passive wave guides, optical switchers, biosensors, masks for lithography, etc. All these ongoing advances since the beginning of the twenty-first century have led to a proper denomination for this new branch of physics: Plasmonics [20]. This new area can be defined as the science and technology of metal optics and nanophotonics, opening up a wide range of practical applications involving manipulation of light at the nanoscale, such as biodetection at the single-molecule level, enhanced optical transmission through sub-wavelength apertures, or high-resolution optical imaging below the diffraction limit [21–23].

The goal of this book is to provide an overview of collective plasmon modes in a gain media. Discussion is focused on the interaction/coupling between a quantum emitter as a rare-earth ion and a metal surface, resulting in the polarization of electrons in the metal (the formation of an SPP) and/or in the modification of the electronic transitions of the emitter. For the latter, a local field enhancement can be observed and sustained as a result of resonance coupling, or an energy transfer may also occur, due to nonresonance coupling between the metallic nanostructures and the optically active medium. Numerous possibilities are thus arising for integrated optics at the nanoscale to counteract the absorption losses in the metal, to enhance luminescence, and to control the polarization and phase of the quantum emitters, in turn enabling exciting applications of SPPs in nanophotonics and potential large-scale applications in commercial optical telecommunications.

1.2   The Dielectric Function of the Free Electron Gas

1.2.1   Drude Model for Free Electron Gas

The description of plasmon effects contributing to propagation in sub-wavelength apertures [24], nanostructured waveguides [25], and nanoantennas [26, 27], and in localized excitations via tunable radiation of metallic nanoparticles (NP) [28, 29] relies on an understanding of the optical response of the medium to the application of external electromagnetic (EM) fields, especially in noble metals, which constitute a basic type of material in plasmonics. For such analysis, classical electrodynamics is a reasonable approach to dealing with the optical properties of electric conductors.

The theoretical treatment adopted for metals begins with the assumption that the free charge density of the carriers presents inside them is greater compared to that of the bound electrons. These free electrons behave like the molecules that constitute a gas delimited by an arbitrary volume, in which electron–electron collision is the only interaction between them. For this reason, the theoretical model used to describe such a class of electrons in metals is the so-called model of a free electron gas, also known as the Drude model [30]. For simplicity, it is easier to consider the equation of motion for only one electron:

$$m\ddot{r} + m\gamma\dot{r} = eE_0 e^{-i\omega t}. \tag{1.1}$$

The following notation has been adopted for this equation. The free electron's properties are the absolute value of elementary charge (e), the effective electron mass (m), and the damping factor (γ), related to the rate collisions between electrons that constitute the gas. Meanwhile, the field's properties are its frequency (ω) and its amplitude (E_0). Although the electron has a negative charge, it is not driven by an attractive force, like in the mass–spring model for instance, because the equation of motion describes only a free electron. For this reason, the right-hand side of Eq. (1.1) is positive. As the field possesses a harmonic dependence in time, it is reasonable to expect that the response of the electron's motion for the applied field will also be harmonic.

From this assumption, we find the solution of Eq. (1.1) to be $r = r_0 e^{-i\omega t}$. Additionally, it is possible to extract further quantities for the velocity and acceleration:

$$\dot{r} = -i\omega r_0 e^{-i\omega t} \tag{1.2}$$

and

$$\ddot{r} = -\omega^2 r_0 e^{-i\omega t} \tag{1.3}$$

Applying the latter results for the equation of motion yields the following amplitude r_0:

$$r_0 = -\frac{e}{m(\omega^2 + i\gamma\omega)} E_0 \qquad (1.4)$$

The amplitude oscillation of the free electron is a complex quantity. It possesses a phase shift caused by the external field and it is influenced by the rate of collisions between the free electron and other electrons present in the gas. We substitute the solution r into Eq. (1.4):

$$r(t) = -\frac{e}{m(\omega^2 + i\gamma\omega)} E(t) \qquad (1.5)$$

The motion of a free electron in metals described by $r(t)$ defines the main feature of conductors: the response of this material to an applied EM field is represented by complex quantities, which have a physical meaning, such as the dielectric function.

1.2.2   Optical Properties of Metals: Drude Model

The application of an external electric field E over a certain material leads to a movement r of its charge carriers. Such displacement generates an electric dipole momentum p inside the material. For noble metals, in which the density of free charge carriers is about 10^{28}, the applied field results in a considerable volumetric distribution of dipole momentum inside the metal, known as the macroscopic polarization $P = n\,p$, where n is the number of charge carriers per unit volume. For the following analysis, we consider noble metals as linear, isotropic, and homogenous materials. This considerable approximation is important when we relate the macroscopic electric field to the polarization from the medium. This assumption produces:

$$D(\omega) = \varepsilon_0 E(\omega) + P(\omega) \qquad (1.6)$$

It is convenient to write all the physical quantities as a function of frequency, ω, instead of time, t, because the aim is to determine the dispersive properties. Here, ε_0 is the vacuum permittivity. The dipole momentum p is defined as:

$$p(\omega) = er(\omega) \qquad (1.7)$$

Substituting this result for the polarization in Eq. (1.6) produces:

$$D(\omega) = \varepsilon_0 E(\omega) + n\,e\,r(\omega)$$

Next, we substitute the result for electron displacement from the Drude model of a free electron gas expressed in Eq. (1.5):

$$D(\omega) = \varepsilon_0 E(\omega) - \frac{n\,e^2}{m(\omega^2 + i\gamma\omega)} E(\omega)$$

$$D(\omega) = \varepsilon_0 \left[1 - \frac{\omega_p^2}{(\omega^2 + i\gamma\omega)} \right] E(\omega).$$

$$(1.8)$$

The quantity between the brackets defines the electric permittivity, or the dielectric function $\varepsilon(\omega)$:

$$\varepsilon(\omega) = 1 - \frac{\omega_p^2}{(\omega^2 + i\gamma\omega)} \qquad (1.9)$$

where $\omega_p^2 = \frac{ne^2}{\varepsilon_0 m}$ is the square of the plasma frequency for a free electron gas. For noble metals, the plasma frequency has values over 10^{16} s^{-1} in optical regime of the EM spectrum, while the collision rate (or damping factor) has values around 10^{14} s^{-1}. The electric permittivity allows us to define the refractive index $\tilde{n} = n(\omega) + i\kappa(\omega) = \sqrt{\varepsilon(\omega)}$. Since the dielectric function is a complex quantity, it is appropriate to write it in terms of both its real and imaginary parts:

$$\mathrm{Re}[\varepsilon_m(\omega)] = 1 - \frac{\omega_p^2}{\omega^2 + \gamma^2} \qquad (1.10)$$

and

$$\mathrm{Im}[\varepsilon_m(\omega)] = \frac{\gamma\omega_p^2}{\omega(\omega^2 + \gamma^2)}. \qquad (1.11)$$

The damping factor from the dielectric function is related to the Fermi velocity v_F and the electron mean free path l (the corresponding distance between collisions of the electrons) by the relation $\gamma = \frac{v_F}{l}$. Furthermore, it is possible to rewrite the properties of the dielectric function in terms of the characteristic time of the free electron, $\tau = \frac{1}{\gamma}$:

$$\mathrm{Re}[\varepsilon_m(\omega)] = 1 - \frac{\omega_p^2 \tau^2}{1 + \omega^2 \tau^2} \qquad (1.12)$$

and

$$\mathrm{Im}[\varepsilon_m(\omega)] = \frac{\omega_p^2 \tau}{\omega(1 + \omega^2 \tau^2)}. \qquad (1.13)$$

The real part of the dielectric function, $\mathrm{Re}[\varepsilon_m(\omega)]$, is associated with the capacity of metal to be polarized by an external electric field, while the imaginary part, Im

$[\varepsilon_m(\omega)]$, is related to energy loss phenomena to which free electrons are subjected, such as the response time for polarizing the metal or losses from absorption of radiation inside the medium, for instance. We note that the real part in Eqs. (1.10) and (1.12) is a negative quantity, since the plasma frequency is greater than the damping factor and field frequency.

A full understanding of the dielectric function is vital for the development of plasmonic devices [31], in order to avoid the considerable losses that can occur in waveguides, resonators, or even solar cells that employ plasmonic effects in their applications. Furthermore, from the results of the dielectric functions presented in Eqs. (1.12) and (1.13), it is important to note that the response of the surface metal to the application of an external electric field depends on time, specifically the characteristic time τ, i.e., there is a delay between the oscillation time of the field, associated with its frequency ω, and τ.

The Drude model for a free electron gas is reliable only for a very low frequency regime ($\omega\tau \ll 1$). Consequently, it is suitable for the microwave and infrared domains. Nevertheless, when the frequency of EM radiation approaches the ultraviolet (UV) or even some parts of the visible regime, the Drude model breaks down. As the majority of metals exhibit a plasma frequency in the UV domain, the response of the electrons to the external field decreases as a result of interband transitions [32]. The divergence between the theoretical model and experimental measurements represented by Palik's database [33] of the real and imaginary parts of the dielectric function in terms of radiation frequency is depicted in the graphs in Fig. 1.1. Results are given for gold (Au), silver (Ag), and aluminum (Al), which are the first metals extensively used for the development of plasmonic devices.

The evidence shows that the Drude model for a free electron gas is incomplete, as it is unable to fully describe the optical properties of the metals shown in Fig. 1.1. Metals like copper and platinum also presented similar problems. The imaginary part diverges greatly from the theoretical model compared to the real part. Since Im $[\varepsilon_m(\omega)]$ is associated with the losses to which free electrons are subjected, such as electron–electron interaction (collisions or Coulomb repulsion), electron–phonon interaction, or even lattice defects due to impurities arising from the deposition process of the metal over a substrate, this can lead to deviations from the model. Furthermore, although metals have a considerable number of free charge carriers, they also possess bound electrons. These bound electrons are responsible for ohmic losses. The absorption of photons, related to the imaginary part of the dielectric function, promotes bound electrons from lower energy levels into higher levels and is the cause of interband transition.

It is clear that the Drude model does not fit well with the experiment, as it was developed before quantum theory. Therefore, to overcome this problem, we must also consider the effects of bound electrons. The next section will investigate this issue.

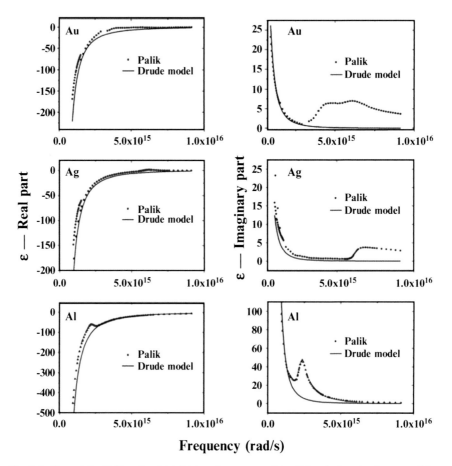

Fig. 1.1 Plots of the (*left*) real and (*right*) imaginary parts of the dielectric function in terms of the radiation frequency for (*top*) gold, (*middle*) silver, and (*bottom*) aluminum. The curves that represent the Drude model for the dielectric function are depicted as *solid blue lines*. We used the expressions of Eqs. (1.10) and (1.11) to plot the theoretical curves. The values of the plasma frequency are: $\omega_p^2 = 1.37 \times 10^{16}$ rad/s (Au/Ag) and $\omega_p^2 = 2.24 \times 10^{16}$ rad/s (Al). For the damping factor, the values are: $\gamma = 1.07 \times 10^{14}$ rad/s (Au), $\gamma = 5.56 \times 10^{13}$ rad/s (Ag), and $\gamma = 1.73 \times 10^{14}$ rad/s (Al). The dots are the experimental measurements of Palik extracted from [33]

1.2.3 Optical Properties of Metals: Interband Transitions

Bound electrons are subject to a restoring force, like the mass–spring system. Therefore, it is convenient to write the equation of motion in Eq. (1.1) with an additional term:

$$m\ddot{r} + m\gamma\dot{r} + kr = -eE_0e^{-i\omega t}, \qquad (1.14)$$

where k is a spring constant related to the restoring force. We note that, since the electron is bound, the external force, which is related to the electric field by the right-hand term of the equation of motion, has a negative value. By dividing both sides of the equation by the mass m and using the same approximation for the free electron's displacement, the amplitude of the oscillating bound electron is given by:

$$r_0 = \frac{-(e/m)}{\left(\omega_0^2 - \omega^2\right) - i\gamma\omega} E_0, \qquad (1.15)$$

where $\omega_0 = \sqrt{\frac{k}{m}}$ corresponds to the natural frequency of oscillation of the bound electron. From the relationship between electric displacement, the electric field, and polarization, we obtain the following constitutive relation:

$$D(\omega) = \varepsilon_0 \left[1 + \frac{\omega_p^{2'}}{\left(\omega_0^2 - \omega^2\right) - i\gamma\omega} \right] E(\omega). \qquad (1.16)$$

The quantity inside the brackets is known as the electric permittivity, $\varepsilon(\omega)$, (or the dielectric function) for bound electrons:

$$\varepsilon(\omega) = 1 + \frac{\omega_p^{2'}}{\left(\omega_0^2 - \omega^2\right) - i\gamma\omega}. \qquad (1.17)$$

We note that this dielectric function is different from that expressed by Eq. (1.9). The presence of the natural frequency ω_0 indicates that the electron is bound. Moreover, the plasma frequency $\omega_p^{2'} = \frac{ne^2}{\varepsilon_0 m^*}$ is quite different from that for free electrons, since the effective mass of a bound electron is not the same as that in the first case. For this reason, we change the notation for the mass $(m \rightarrow m^*)$. From Eq. (1.17), it is possible to identify the real and imaginary parts, since the dielectric function is also complex:

$$\mathrm{Re}[\varepsilon_m(\omega)] = 1 + \frac{\omega_p^{2'}\left(\omega_0^2 - \omega^2\right)}{\left(\omega_0^2 - \omega^2\right)^2 + \gamma^2\omega^2}. \qquad (1.18)$$

and

$$\mathrm{Im}[\varepsilon_m(\omega)] = \frac{\gamma\omega\,\omega_p^{2'}}{\left(\omega_0^2 - \omega^2\right)^2 + \gamma^2\omega^2}. \qquad (1.19)$$

The natural frequency (ω_0) also has an important physical meaning. It defines the resonance frequency that the metal can have with the EM spectrum. In the quantum mechanical picture, this specific frequency is proportional to the energy of the photon absorbed by the bound electrons during their transition from lower energy levels to the conduction band. In general, metals possess more than one resonance frequency, especially in the visible range. This is because not all bound

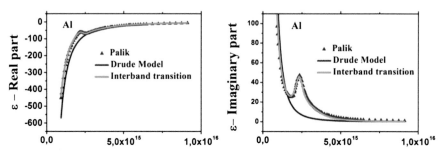

Fig. 1.2 The (*left*) real and (*right*) imaginary parts of the dielectric function for aluminum plotted in the visible range of the EM spectrum. There are three situations depicted in each graph: (1) the *blue* (real) and *red* (imaginary) *triangles* denote experimental data for the dielectric function; (2) the curves from the Drude model for free electrons that do not fit well with the experimental case, and (3) the theoretical curves that consider the interband transitions due to bound electrons (*solid green lines*). Clearly, the final approximation from Eqs. (1.18) and (1.19) is more reliable when compared to the Drude model. The *dots* are the experimental measurements of Palik extracted from [33]

electrons begin the transition from the same level. The energy needed for a photon to be absorbed by the electron on state 1 s is different from that of one on state 2 s, for instance. For this reason, even the model for bound electrons is limited in its description of the dielectric function. A simple case in which a metal has only two resonance frequencies is depicted in Fig. 1.2. The real and imaginary parts of the dielectric function are plotted for the case of Al.

To plot the curves corresponding to the interband transition, we have expanded the dielectric functions into three components: one related to the case in which there is no resonance, i.e., for free electrons, equivalent to the Drude model case; and the two remaining components are associated with the interband transition with their respective resonance frequencies. The expansion for each dielectric function is written in the following equations:

$$\mathrm{Re}[\varepsilon_m(\omega)] = 1 - f_1\frac{\omega_p^2}{\omega^2 + \gamma^2} + f_2\frac{(\omega_0^2 - \omega^2)\omega_p^2}{(\omega_0^2 - \omega^2)^2 + \gamma_0^2\omega^2}$$
$$+ f_3\frac{(\omega_1^2 - \omega^2)\omega_p^2}{(\omega_1^2 - \omega^2)^2 + \gamma_1^2\omega^2}. \tag{1.20}$$

and

$$\mathrm{Im}[\varepsilon_m(\omega)] = f_1\frac{\gamma\omega_p^2}{\omega(\omega^2 + \gamma^2)} + f_2\frac{\gamma_0\omega\omega_p^2}{(\omega_0^2 - \omega^2)^2 + \gamma_0^2\omega^2}$$
$$+ f_3\frac{\gamma_1\omega\omega_p^2}{(\omega_1^2 - \omega^2)^2 + \gamma_1^2\omega^2}. \tag{1.21}$$

Table 1.1 Parameters used to plot the curves for Al

Parameters of the plotting for aluminum	
ω_p	2.24×10^{16} rad/s
ω_0	2.356×10^{15} rad/s
ω_1	3.141×10^{15} rad/s
γ	1.733×10^{14} rad/s
γ_0	6.31×10^{14} rad/s
γ_1	1.8×10^{15} rad/s
f_1	0.8
$f_2 = f_3$	0.1

Terms proportional to the coefficient f_1 in both parts of the dielectric function do not display resonance, similar to the cases defined in Eqs. (1.10) and (1.11), in agreement with the Drude model. However, terms proportional to the coefficients f_2 and f_3 are related to the resonance frequencies of the bound electrons. These three coefficients are statistical weights that indicate which state the electrons are in. For metals, the probability to find a free electron state is greater than that for the bound state. In general, $f_1 > f_2, f_3$ when such an analysis is carried out. Furthermore, to plot the curves for Al, we have approximated the plasma frequency of the bound electrons as equal to that of the free electrons ($\omega_p' \approx \omega_p$), since these quantities are quite similar. The specifications of the parameters used to plot the curves of Al are shown in Table 1.1.

The corresponding energies of the photons absorbed by the bound electrons are found using the resonance frequencies ω_0 and ω_1. We have obtained the following values: $E_0 = \hbar\omega_0 \cong 2.48$ eV and $E_1 = \hbar\omega_1 \cong 3.31$ eV. Additionally, the wavelengths associated with these resonances are: $\lambda_0 = \frac{hc}{E_0} \cong 500$ nm and $\lambda_1 = \frac{hc}{E_1} \cong 375$ nm, where $hc = 1240$ eV. nm and h and $\hbar$ are the Planck and reduced Planck constants, respectively. We note that these wavelengths correspond to the visible and UV domains, respectively. These regions of the EM spectrum are exactly the ranges in which the Drude model fails to describe the dielectric function. An accurate theoretical and experimental analysis for other metals, similar to that carried out for Al, can be found in the literature [32, 33].

1.3   Surface Plasmons, Polaritons, and the Gain Medium

In the previous section, we examined the optical properties of metals and their response to an incident external EM wave. They present peculiar properties, such as negative values for the electric permittivity (the real part) and high reflectivity for the visible region of the EM spectrum. Such properties arise from the conductive nature of metals, which confers a damping effect on the propagation of the EM wave into the material. This feature is critical to confining and controlling light for the development of plasmonic devices, since the damping is related to the losses that these materials can exhibit. To overcome this problem, the introduction of a second medium is necessary to compensate for these losses. This section is dedicated to describing the optical response of gain media to improve the capacity of light confinement.

Dielectrics are the best known type of material that can be combined with metals for the coupling of plasmons, as we will see in Chap. 2. For now, our goal is to obtain the optical response for dielectrics using a classical treatment similar to that developed in the preceding section. The external electric field exhibits a time harmonic dependence, $E(t) = E_0 e^{-i\omega t}$; as a consequence, the electron responds similarly and therefore its vector position is $r(t) = r_0 e^{-i\omega t}$. The equation of motion for one electron is given by Eq. (1.14). This equation is similar to that for one electron in metal. The only exception is the term proportional to the constant k associated with a restoring force; see Eq. (1.1). For dielectrics, a considerable proportion of the electrons are bound, unlike metals in which they are free, and the term kr can be neglected. Therefore, the solution $r(t)$ for the equation of motion is Eq. (1.15).

As with metals, dielectrics are also polarized by the external EM field. A volumetric distribution for all dipoles produces the polarization $P = np$. The polarization is given by the following expression:

$$P = -ner = \frac{\frac{ne^2}{m}}{\left(\omega_0^2 - \omega^2\right) - i\omega\gamma} E(t) = \varepsilon_0 \chi(\omega) E(t), \tag{1.22}$$

where $\chi(\omega)$ is the electric susceptibility, expressed by:

$$\chi(\omega) = \frac{\omega_p^2}{\left(\omega_0^2 - \omega^2\right) - i\omega\gamma}. \tag{1.23}$$

From the relationship between D and E (see Eq. (1.6)), we obtain:

$$\varepsilon(\omega) = \varepsilon_0(1 + \chi(\omega)) = \varepsilon_0 + \frac{\varepsilon_0 \omega_p^2}{\left(\omega_0^2 - \omega^2\right) - i\omega\gamma}. \tag{1.24}$$

This is the electric permittivity for the dielectrics. The main difference between metals and dielectrics is the presence of a resonance frequency ω_0 for the latter; ω_0 also indicates that the electrons are bound. Furthermore, for dielectrics, the real part of the permittivity is a real quantity, as follows:

$$\mathrm{Re}[\varepsilon(\omega)] = \varepsilon_0 + \frac{\varepsilon_0 \omega_p^2 \left(\omega_0^2 - \omega^2\right)}{\left(\omega_0^2 - \omega^2\right)^2 + \gamma^2 \omega^2} \tag{1.25}$$

and

$$\text{Im}[\varepsilon(\omega)] = \frac{\varepsilon_0 \omega_{\text{p}}^2 \gamma \, \omega}{\left(\omega_0^2 - \omega^2\right)^2 + \gamma^2 \omega^2}. \tag{1.26}$$

It is worth noting that the real part of the permittivity indicates the capacity for the dielectric to be polarized, and that the imaginary part is associated with the absorption of the material. For frequencies close to the resonance frequency ($\omega \approx \omega_0$), the absorption of the dielectric is maximum, as Eq. (1.26) suggests. However, there is a decrease of the real part for this frequency region; this phenomenon is known as anomalous dispersion [34].

The model developed to obtain expressions for the permittivity is the so-called Drude–Lorentz model, or simply named the oscillator model. It complements the Drude model of the free electron gas for the optical response of metals. These analogies with mechanical systems are good approximations for explaining the polarization and absorption processes and display a great applicability to the experimental data. Nevertheless, for more realistic dielectric materials, there are other features that they may present compared with the simple case of the Drude–Lorentz model. One of the most remarkable characteristics that differs from the initial model is the presence of additional resonance frequencies in the expression of the permittivity [21, 22], e.g., quantum emitters (QE) within the material. These new resonances are related to the complementary vibrational modes of the electrons resulting in defects in the lattice. When we have rare-earth ions (REI) in the substrate, there are resonances associated with the transitions between electronic levels of the REIs. Therefore, a reliable expression for the permittivity that possesses these quantum properties is:

$$\varepsilon(\omega) = \varepsilon_0 + \varepsilon_0 \sum_{j>i} \frac{f_{ji}\left(N_i - N_j\right)e^2 / m\varepsilon_0}{\left(\omega_{ji}^2 - \omega^2\right) - i\omega\gamma_{ji}}. \tag{1.27}$$

In this expression, ω_{ji} denotes the frequency related to the transition of the electrons between the energy levels E_j and E_i. As is typical for the quantum mechanical case, this new resonance frequency is written as $\omega_{ji} = \frac{\left(E_j - E_i\right)}{\hbar}$. Meanwhile, f_{ji} is the sum of the oscillator strength, proportional to the probability of the transition levels, while γ_{ji} is the corresponding damping factor related to this transition. Finally, N_i and N_j correspond to the electronic populations.

For simplicity, we will consider a two-level atomic system. From Eq. (1.27), the permittivity that describes such a system is written as:

$$\varepsilon(\omega) = \varepsilon_0 + \varepsilon_0 \frac{f\omega_{\text{p}}^2}{\left(\omega_{21}^2 - \omega^2\right) - i\omega\gamma_{21}}, \tag{1.28}$$

a

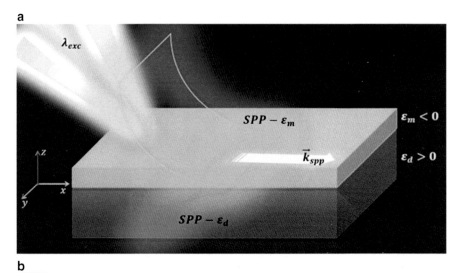

b

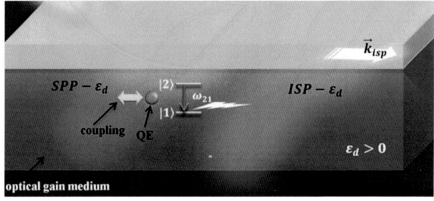

Fig. 1.3 (**a**) A light beam, with wavelength λ_{exc}, creating an SPP on both faces of a metallic surface, with $\varepsilon_m < 0$. (**b**) Dipoles near a metal–dielectric interface. Within the dielectric, there is a QE of two energy levels near the metallic surface (an optical gain medium)

where:

$$f = f_{21} \frac{N_1 - N_2}{N_1 + N_2} \qquad (1.29)$$

and

$$\omega_p^2 = \frac{(N_1 + N_2)e^2}{m\varepsilon_0}. \qquad (1.30)$$

An example of this system is depicted in Fig. 1.3. The origin of the gain medium is related to the number of electrons that each level can contain. For ordinary materials, the population of electrons that lie in the fundamental level is greater than the number of electrons in the excited level ($N_1 > N_2$). For this case, the permittivity is similar to the case with only one resonance frequency. However, when there is an inversion in the number of electrons, i.e., $N_1 < N_2$ (this case is only for a single QE with two energy levels), the oscillator strength becomes negative ($f < 0$). As a consequence, the imaginary part of the permittivity, initially associated with the absorption of EM radiation, is related to the gain medium in this new case, because it is also negative.

1.3.1 Surface Plasmon Polariton-Coupled Emission

Surface plasmon polaritons (SPP), often referred to simply as surface plasmons (SP), are EM waves that result from the interaction of light with a metallic surface, which stimulates collective oscillations of conduction electrons that propagate in a direction parallel to the metal–dielectric interface. The propagation modes can be short- or long-range (from tens to hundreds of micrometers) and depend intrinsically on the nature of the surface of the metal and/or the dielectric. Nevertheless, SPPs only occur for transverse magnetic (TM) polarization [20]. When the frequency of incident light matches the natural frequency of the SPPs, surface plasmon resonance (SPR) occurs. Consequently, SPR localized in a nanostructure is called localized surface plasmon resonance (LSPR) and it possesses a certain normal mode. The topology of the metallic surface and the bordering dielectric determines the characteristics of the diverse SPP propagation modes that occur, such as on flat and curved surfaces and on single or multiple surfaces, as well as the SPP modes of complex NP arrays and metallic nanostructures. These phenomena (SPP and LSPR) are also extremely sensitive to subtle changes in the dielectric (e.g., refractive index) or to the geometry and size of the metallic nanostructures. These phenomena are in fact the basis of a number of applications, including devices that enhance the local field; extraordinary optical transmission (EOT); nanoantennas for optical probes in chemistry, biology, and medicine (exciting locally with sub-wavelength resolution); plasmonic photothermal therapy; surface-enhanced Raman scattering; fluorescence enhancement; the implementation of sub-wavelength waveguides; confinement of light in a nanoscale region; and the development of efficient solar cells. In these applications, SPPs open rich possibilities for plasmonic nanocircuits, in which light would carry information along complex paths over distances of more than 100 μm and passing through many processing steps.

The collective electron oscillations of SPs can be interpreted as a displacement of the center of mass of all electrons in the surface against the positively charged background of the surface (i.e., the atomic cores); see Eqs. (1.1) and (1.14).

One important parameter is the incident angle of the impinging light, since this defines the lateral component, i.e., the wave vector for free space, $k_0 \left(= \frac{\omega}{c} = \frac{2\pi}{\lambda}\right)$, parallel to the metallic surface, matches the wave vector for the SPP, $k_{spp} \left(= k_0 \sqrt{\frac{\varepsilon_d \varepsilon_m}{\varepsilon_d + \varepsilon_m}}\right.$

(the permittivities of the metal, ε_m, and the dielectric material, ε_d, must have opposite signs if SPPs are to be possible at such an interface). Then, the SPPs are in resonance with the frequency of the incident light. Under these conditions, an EM field efficiently couples to the SPP, although it is highly attenuated as a result of the light reflection, as shown in Fig. 1.3.

Furthermore, the SPP can create greatly enhanced evanescent fields penetrating the dielectric above the metal surface, up to several micrometers into the substrate, as shown in Fig. 1.3a. Within this distance, r, let us assume the existence of QEs in the dielectric, which can be excited by the evanescent field, creating an EM field that may strongly interact with free charges in the metallic film, creating induced surface plasmons (ISP) at the metal–dielectric interface. The frequency of these ISPs corresponds to the emission frequency of the QE, as in Fig. 1.3b; consequently, their emissions have the same spectral shape and are highly polarized as a result of induced polarization from the SPPs in the metallic films, and they in turn radiate into the dielectric (transparent material). The second consequence of the interaction between the surface charges and the EM field is that, in contrast to the propagating nature of SPPs along the surface, the field perpendicular to the surface decays exponentially with distance from the surface, as shown in Fig. 1.3a.

As mentioned above, the coupling of light with conduction electrons allows us to break the diffraction limit for the localization of light into sub-wavelength dimensions, enabling strong field enhancements and opening up new opportunities for controlling light confinement on the nanoscale.

The physical properties of SPPs can be studied through the Maxwell's equations in a dielectric medium:

$$\nabla \cdot \boldsymbol{D} = \rho_{ext}$$

$$\nabla \cdot \boldsymbol{B} = 0$$

$$\nabla \times \boldsymbol{E} = -\frac{\partial \boldsymbol{B}}{\partial t} 0 \qquad (1.31)$$

$$\nabla \times \boldsymbol{H} = \boldsymbol{J}_{ext} + \frac{\partial \boldsymbol{D}}{\partial t}$$

where $\boldsymbol{H}$ is the magnetic field and $\boldsymbol{B}$ is the magnetic induction, with the external charge and current densities ρ_{ext} and $\boldsymbol{J}_{ext}$, respectively. We apply the boundary conditions for $\boldsymbol{E}$ and $\boldsymbol{H}$ at the metal–dielectric interface, both with different EM properties. However, both components, $\boldsymbol{E}$ and $\boldsymbol{H}$, are tangent to the surface and must be continuous across the boundary. Additionally, let us represent the QEs as oscillating dipoles. Thus, the EM field of these QEs can be expressed as integrals over plane waves interacting with the metallic film/structure, resulting in a directional emission that can be theoretically investigated. Therefore, the integrals for the electric and magnetic fields at an arbitrary position in the metallic structure, from a dipole normal $\boldsymbol{E}_{QE}$ and $\boldsymbol{H}_{QE}$ to the plane are [21]:

$$E_{QE} = \frac{i\mu_0 c^2 k^3 p_{QE}}{4\pi n n_d} \int_0^\infty \frac{n_\rho^2}{\sqrt{n_d^2 - n_\rho^2}} \{in_z J_1(kn_\rho\rho)[ae^{ikn_z z} - be^{-ikn_z z}]\hat{\rho}_r$$
$$-n_\rho J_0(kn_\rho\rho)[ae^{ikn_z z} + be^{-ikn_z z}]\hat{z}\}dn_\rho \quad (1.32)$$

and

$$H_{QE} = -\frac{ck^3 p_{QE}}{4\pi} \int_0^\infty \frac{n_\rho^2}{\sqrt{n_d^2 - n_\rho^2}} \{J_1(kn_\rho\rho)[ae^{ikn_z z} + be^{-ikn_z z}]\hat{\varphi}_r\}dn_\rho. \quad (1.33)$$

Here, n_d is the refractive index of a dielectric containing the QE, $n_\rho = k_\rho/k$, where k_ρ is the in-plane component of the wave vector; $n_z = k_z/k$, where k_z is the z-component of the wave vector; c is the speed of light in vacuum; μ_0 is the vacuum permeability; a and b are the electric field components for the forward and backward EM components, respectively; J_m are the Bessel functions of order m; $\hat{\varphi}_r$ is the unit vector in the azimuthal direction; $\rho = \sqrt{x^2 + y^2}$; x, y, and z are the Cartesian coordinates; and p_{QE} is the radiating dipole moment of the QE. The radiated power of a dipole in a dispersive dielectric medium is proportional to the refractive index, and in the case of REIs, its emission is very sensitive to the local environment (see Sect. 1.5).

Let us consider a single QE located at a distance r and represented by a two-level system with a transition frequency ω_{qe} with emission probability q. Then, we excite the QE of the SPP at the peak of their emission spectrum; here, the emission rate is written as $\gamma_{em} = q\gamma_{exc}$, where γ_{exc} is the excitation rate $(=|p \cdot E|^2)$ and depends on the local excitation field $E(r, \omega_{spp})$. Therefore, the probability of transition can be expressed as (using Fermi's golden rule):

$$\gamma_{em} = \frac{2\omega_{qe}}{3\hbar\varepsilon_d} |p_{qe}|^2 g(r, \omega_{qe}). \quad (1.34)$$

Here, g denotes the EM density of states, which is strongly modified by the existence of the SPP [35], causing a reduction or increase in the lifetime of the QEs when these are placed close to a metal surface ($\leq r$) [36–38].

Under this scenario, a collective electronic excitation in the surface of a solid plays an essential role in many dynamic surface phenomena and processes. For example, when an SPP accompanies large density oscillations localized near the surface and when coupled with a QE, quantum oscillations can be produced, since the SPP is a function of ω, as in Eq. (1.34). In this situation, quantum oscillations have been observed in the transition temperature of superconductors [39], electron–phonon coupling [40], and others. These oscillations are attributable to the quantization of electronic states normal to the surface; see Sect. 2.2.1, Chap. 2. Surface plasmon-coupled emissions can be easily implemented, e.g., in fluorescence microscopy, to facilitate the signal collection of a single QE, enabling us to develop novel approaches for simpler design of fluorescence-based detection devices.

1.4 Quantum Emitters

Single quantum emitters (QE) have become a common tool for the development of new light sources, such as lasers, LEDs, and single-photon sources, for electronic nanodevices. A single QE is also important in chemistry and life sciences, where they can act as nanoscopic probes and labels. In particular, in the field of quantum information, QEs are utilized as sources of single photons or as stationary quantum bits [41].

As particles become smaller, the laws of quantum mechanics become more important in the study of the particle's interaction with light. In this framework, continuous scattering and absorption of light will be replaced by resonance inter-actions if the photon energy matches the energy difference of the discrete internal (electronic) energy levels. In atoms, ions, molecules, and nanoparticles (as in REIs, semiconductor nanocrystals, and quantum dots), and other quantum-confined sys-tems, these resonances are found at optical frequencies. Because of their resonance characteristics, the light–matter interaction can often be treated as a two-level quantum system, with only those two (electronic) levels whose difference in energy is close to the interacting photon energy $\hbar\omega$. In this manner, a QE is generally defined as a quantum system that is capable of radiative optical transitions. When observing the spontaneous decay of a single excited QE, the emission of a single photon is expected. When suppressing nonradiative decay mechanisms, they can principally act as 100 % efficient single-photon sources. Therefore, transitions of two electronic levels in a QE are ideal systems for generation of a single photon. In recent experiments, single atoms or single ions were coupled to high-finesse cavities to control the emission time and mode of the photon [42], since the emission of an isolated single atom is characterized by a high mode purity, which is important for experiments based on two-photon interference. Atom decay cas-cades can also be used for entangled photon generation [43]. A QE can be excited optically (e.g., into excited states or into the continuum above the bandgap) from the ground state, and then a spontaneous single-photon emission can occur.

The absorption of light by a quantum system can be characterized by a frequency-dependent absorption cross section. In this case, the absorption cross section of the system is given by:

$$\sigma(\omega) = \frac{\omega}{3}\sqrt{\frac{\mu_0}{\varepsilon_0}}\frac{\mathrm{Im}[\alpha(\omega)]}{n_\mathrm{d}(\omega)}, \tag{1.35}$$

where $\alpha(\omega)$ is the polarizability. For instance, for a quantum dot (QD), the value of σ is correspondingly higher because of the QD's increased geometric size, suggesting that every photon passing within the QD's area is absorbed by the QD. Of course, this is a naive picture that from the point of view of quantum mechanics cannot be true, owing to the uncertainty relation, which does not allow the photon to be localized. The resonance characteristics of the interaction lead to the typical phase relation between the driving field and the dipole oscillator response (from the QE), which can change from "in phase" for frequencies far below the resonance to "antiphase" for frequencies far above the resonance. Exactly at resonance, a phase shift of $\lambda/2$ occurs between the excitation and the dipole of the QE.

On the other hand, the radiative relaxation of the QE is called fluorescence. Relaxation can also occur nonradiatively, as vibrations or collisions ultimately lead to energy loss via heat or phonons. The quantum yield of the radiative decay rate k_r and the total radiative decay rate $(k_r + k_{nr})$ is given by:

$$\eta = \frac{k_r}{k_r + k_{nr}}, \tag{1.36}$$

where k_{nr} is the nonradiative decay rate. The emission spectrum consists of a sum of "Lorentzians" [43], also called vibrational progression, corresponding to the different decay pathways into ground-state levels.

Due to non-negligible spin-orbit coupling in QEs (e.g., REIs or heavy metals), there is a finite torque acting on the spin of the electron in the excited state. Therefore, there is a small but significant probability that the spin of the excited electron is reversed. Spin 1 has three possible orientations in an external magnetic field, leading to a triplet of eigenstates, also called a triplet state as opposed to a singlet state for spin 0. The energy of the electron in the triplet state is usually reduced with respect to the singlet's excited state because the exchange interaction between parallel spins increases the average distance between the electrons in accordance with Hund's rule [56]. Consequently, the increased average distance leads to a lowering of their Coulombic repulsion. Once a QE undergoes an intersystem crossing into the triplet state, it may decay into a singlet ground state. However, this is a forbidden spin transition ($\Delta S = 1$) [56]. Triplet states therefore have an extremely long lifetime on the order of milliseconds.

On the other hand, we can represent a QE as a classical dipole p, a point-like source current located at r_0. Therefore, the power dissipated by a time harmonic system is:

$$P = \frac{\omega}{2} \mathrm{Im}[p^* \cdot E(r_0)]. \tag{1.37}$$

Therefore, the power dissipation can be expressed in terms of the dipole moment of the QE, which is the sum of the radiated power, P_{rad}, and the power dissipated into heat and other channels, P_{loss}.

1.5 The Optical Properties of Rare-Earth Ions

In the periodic table of elements, the group of rare-earth ions (REI) is composed of the lanthanide ions, which extends from lanthanum (La) ($Z = 57$) to lutetium (Lu) ($Z = 71$), including also scandium (Sc) ($Z = 21$) and yttrium (Y) ($Z = 39$),

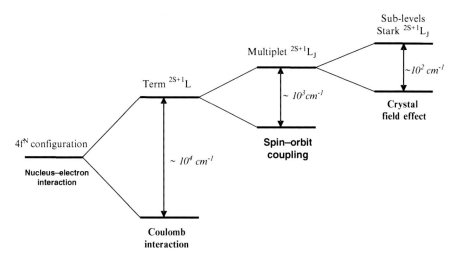

Fig. 1.4 Increasing degeneracy of the energy levels of an REI as a function of the type of interaction

due to their close electronic structures and chemical properties. These chemical elements are often found in the form of trivalent ions, RE^{3+}, with an electronic configuration of $[Xe]\,4f^N$, where N is the number of electrons in the f electron shell. The REIs are well known not only for their magnetic properties (e.g., the neodymium (Nd)-based magnet), but also for their unique luminescent properties, giving them numerous applications in optics and photonics. The unique properties of these ions arise from the low radial expansion of their 4f electronic orbitals. These orbitals are screened by the external 5s and 5p electron shells. Their valence electrons are therefore less sensitive to their chemical surroundings, while transitions between energy levels into the incomplete 4f electron shell usually result in sharp absorption and emission bands ranging from the UV to the middle infrared (MIR). Compared to other optically active ions like metal transition ions, the emission and absorption wavelengths of REIs are relatively insensitive to the host material while their excited state lifetimes and quantum efficiencies are typically longer and higher, respectively. For a comprehensive description of the discovery of REIs, the reader is invited to see refs. [44–48].

The optical (absorption and emission) spectra of an REI are characteristic of the electronic transitions occurring between levels within the $4f^N$ shell. The spectral position of these levels results from the combination of different interactions: electron–nucleus interaction, Coulomb repulsion forces between electrons, spin–orbit coupling, and crystal field interaction. The increasing degeneracy of the energy levels of the REI as a result of these perturbations and the order of magnitude of the energies involved are schematically represented in Fig. 1.4.

The surroundings of the REI influence not only the splitting of the $^{2S+1}L_J$ levels via the crystal field, but also the probabilities of transitions between those levels. Thus, the absorption and emission cross sections may differ according to the

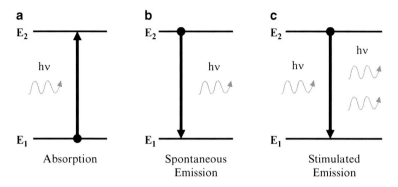

Fig. 1.5 Possible transition paths between two energy levels

chemical nature of the matrix (e.g., oxide, fluoride, or chalcogenide), its linear refractive index, and the ionic/covalent characteristics of its chemical bonding within the material.

In the absorption process for an REI contained within a material, incident photons with an energy $E = h\upsilon$ close to the energy difference ΔE between the fundamental and excited levels of the REI are absorbed, resulting in the excitation transition. In the absence of external electric field, the excited REIs relax to their fundamental states according to two possible processes: one is radiative with the emission of light and the other one is nonradiative with the emission of a phonon (via lattice vibration or heat dissipation).

First, the emission process can be spontaneous or stimulated (Fig. 1.5). In spontaneous emission, the excited ions lose the absorbed energy through light emission by emitting a photon of energy $h\upsilon = E_2 - E_1$, as presented in Fig. 1.5b. In stimulated emission, which is the light amplification phenomenon exploited in lasers, the incident photon may provoke the relaxation (de-excitation) of an electron in the E_2 state by the emission of a photon with the same characteristics (wavelength, phase, direction, and polarization) as the incident photon. Stimulated emission causes the duplication of light and requires an inversion of population between levels E_2 and E_1 to be possible, as described in Fig. 1.5c.

The excited ions can also relax by dissipating energy into the lattice, by creating phonons. If the excess energy is higher than the maximum phonon energy of the host material, then several phonons can be created simultaneously, creating a multiphonon relaxation. It is thought that phonons of higher energy are created in the relaxation process to limit the number involved in the transition. This phenomenon can eventually lead to the heating of the host material.

Finally, the phonon energy of the matrix itself, which is defined by the overall vibrations of its constitutive elements, also plays a key role in the emission properties of the REI. The phonons of the host network may provoke the relaxation of the levels of excited states and consequently lower the quantum emission efficiency.

For further development and a more detailed description of the spectroscopic properties of REIs for optical applications, the reader is strongly invited to see the reference book edited by Liu and Jacquier [45].

1.5.1   Excitation of Localized 4f States

As mentioned in the previous section, the screening of 4f electronic orbitals by 5s and 5p orbitals in an REI makes them much less sensitive to their chemical surroundings than other optically active elements, such as transition metal ions. This behavior is also exploited by using REIs as probes in solids. Indeed, as the crystal environment constitutes a small perturbation on the atomic and solid-state energy levels, the spectroscopic properties of REIs can be interpreted by considering them as free ions.

Let us use the Russell–Saunders notation for the energy levels, $^{2S+1}L_J$, which takes into account spin–spin coupling, orbit–orbit coupling, and spin–orbit coupling. The energy levels of a free REI are typically interpreted by considering only interactions between the 4f electrons themselves. Since all the other electronic shells are spherically symmetric, their effect on all the terms of a configuration is the same at first order, and therefore they do not contribute significantly to the relative positions of the 4f energy levels. We can then write the following expression:

$$H = -\frac{\hbar^2}{2m}\sum_{i=1}^{N}\Delta_i - \sum_{i=1}^{N}\frac{Z^*e^2}{r_i} + \underbrace{\sum_{i<j}^{N}\frac{e^2}{r_{ij}}}_{H_C} + \underbrace{\sum_{i=1}^{N}\zeta(r_i)s_i\cdot l_i}_{H_{SO}} + V_{EF}, \quad (1.38)$$

where $N = 1, \ldots, 14$ is the number of 4f electrons, Z^*e is the screened charge of the nucleus because we have neglected the closed electronic shelves, V_{EF} is a potential modeling the interaction of the ion with the electromagnetic field, and $\zeta(r_i)$ is the spin–orbit coupling function [46] given by:

$$\zeta(r_i) = \frac{\hbar^2}{2m^2c^2r_i}\frac{dU(r_i)}{dr_i}, \quad (1.39)$$

where $U(r_i)$ is the potential in which the electron i is moving. The first two terms of the Hamiltonian, Eq. (1.38), are spherically symmetric and therefore do not remove any one of the degeneracies within the configuration of the 4f electrons. They can be thus neglected here. The Coulomb interaction of the 4f electrons (H_C) and their spin–orbit interaction (H_{SO}) are responsible for the energy level structure of the electronic configurations in 4f, as shown in Fig. 1.4. For more details about the matrix element calculations of H_C and H_{SO}, the reader is invited to read the major references [47, 48].

The crystal field interaction in REIs is weak, as we will briefly discuss in this section. It is also well known that the $4f$ wave functions do not extend very far beyond the $5s$ and $5p$ shells. Any physical quantities of an REI-containing solid that depend on the overlap of $4f$ wave functions with those of a neighboring ion are thus expected to be small. Next, we will discuss crystal field splitting and conclude this subsection with the intensities of radiative optical transitions.

1.5.1.1 Trivalent Ions in a Static Crystal Field

As mentioned previously, the $4f$ shell of REIs is unfilled, therefore making its charge distribution nonspherical. An REI placed into a crystal will experience an inhomogeneous electrostatic field produced by the charge distribution within the crystal: this is the crystal field. This crystal field distorts the closed shells of the REI, thus removing a certain degree of degeneracy in the $4f$ levels of the free ion and producing a major modification of the energy levels as a function of the crystal symmetry; see Eq. (1.38). Additional transitions originating from the existence of the excited levels of the crystal field from the ground term may then be observed in the emission spectrum, which thus appears more complex. On the other hand, such a spectrum can also be used to determine the crystal field energies of the ground term [49, 50]. Because, the crystal field interaction corresponds to the interaction between the $4f$ electrons with all the charges of the crystal; see Eqs. (1.40) and (1.41). Except for radial factors, this interaction is a function of two spherical harmonics, Y_{kq}: the first one contains the coordinates of the $4f$ electrons while the other one contains the coordinates of the crystal charge (integrated over the whole crystal). Therefore, we can determine the strength of the crystal potential at the site of the REI. This splitting of the crystal field into Y_{kq} contains a second-order term, as discussed in the next subsection.

1.5.1.2 Crystal Field Splitting

There are two aspects to crystal field splitting: (1) the symmetry, namely, the number of levels into which the J terms of a free iron are split in a crystal field of a given symmetry; and (2) the actual size of the crystal field splitting. To describe qualitatively the crystal field interaction, we can use the point charge model. If we take into account the fact that the crystal field is built up of the spatially extended charge clouds of the individual ions, then these charge clouds can penetrate each other and interact. Hence, we can consider the following elements:

1. The ions are static in the crystal, i.e., we can neglect the lattice vibrations and their effect on the energy levels.
2. The $4f$ electrons of one single REI are representative of those of all the REIs in the crystal and the interaction of $4f$ electrons of adjacent ions is neglected.

3. The crystal consists of extended charge distributions produced by an overlap of the charge distributions of the neighboring ions and the 4f electrons. In addition, a charge transfer between the 4f electrons and the electrons of the ligands can take place. Both contribute then to the crystal field interaction.
4. The 4f electrons of one REI are considered to be independent of each other, which imply that correlation effects do not play any significant role.

Therefore, we can calculate the crystal field potential, $V(r_i, \varphi_i, \theta_i)$, at the site of the 4$f$ electrons and the potential energy of the 4f electrons in this potential as:

$$V(r_i, \varphi_i, \theta_i) = -\sum_{k,q,i} B_{k,q} C_{k,q}(\varphi_i, \theta_i), \qquad (1.40)$$

where i is the summation carried out over all the 4f electrons of the ions and $B_{k,q}$ is the crystal field parameter ($k \leq 6$ for f electrons). This has the form:

$$B_{k,q} = -e \int (-1)^q \rho(R) C_{k,q}(\varphi_i, \theta_i) \frac{r_<^k}{r_>^{k+1}} dt, \qquad (1.41)$$

where $C_{k,q}$ is a tensor operator defined as $C_{k,q} = \sqrt{\frac{4\pi}{2k+1}} Y_{kq}$; $r_<$ and $r_>$ are, respectively, the smaller and larger values of r and r_i; and $\rho(R)$ is the crystal field charge density.

Initially, it is supposed that the contribution of the point charge would be the dominant part of the interaction of the crystal field with the REIs. Then, in Eq. (1.40), the integral over the lattice can be replaced by a sum over all lattice points and $r_<^k$ can be replaced by r_i^k. The latter replacement can be performed so long as the charge distribution of the crystal does not enter that of the 4f electrons (as long as $r_i < R$), which implies that the potential acting on the 4f electrons obeys the Laplace equation ($\Delta\Phi(r, \varphi, \theta) = 0$) at the position of the 4f electrons.

The 4f^N configuration is composed of a number of states where the quantum numbers (L, S, J, and another arbitrary choice) define the terms of the configuration, all of them being degenerated in the central-field approximation. Then the spin–orbit interaction is found to be the strongest of the magnetic interactions in the ladder, as presented in Fig. 1.4 (partially resulting from the optical properties of REIs). This spin–orbit interaction in Eq. (1.39) increases degeneracy of the J and splits the LS terms into J levels.

The amount of splitting and or shifting is called the Stark splitting or the Stark shift, and the resulting states are called Stark components (of the parent manifold). The even-k terms in the expansion split the J multiplets of the free ion into Stark components generally separated by 10–100 cm^{-1}. The ion–lattice interaction can mix multiplets with different J values (J mixing), although it usually remains a good quantum number. The odd-k terms admit higher lying states of opposite parity (e.g., 4$f^N 15d^1$) into the 4f^N configuration. This admixture does not affect the positions of the energy levels, but it has a very important effect on the strengths of the optical transitions between levels.

The intra-4f transitions are parity forbidden and are partially allowed by crystal field interactions mixing opposite parity wave functions, resulting in long luminescence lifetimes (in the microsecond or millisecond range) and narrow line widths. Through the selection of a particular REI, we can have an appropriate emission band across the visible or infrared region. Figure 1.6 presents the energy level diagrams for isolated RE^{3+} ions.

The 4f electrons are well shielded from the chemical environment and therefore have almost retained their atomic characteristics. Nevertheless, for a number of REIs, broad emission bands are also known. Relevant examples are the Eu^{2+} and Ce^{3+} ions, where the emission comes from 5d to 4f optical transitions. As the involved electrons participate in the chemical bonding, the d–f emission spectra consist of broad bands. These types of transition are allowed and are consequently very fast (a few microseconds or less).

1.5.1.3   Radiative Transition

As the terms shown in Eq. (1.38) are time-dependent, they do not lead to stationary states for the system. They are thus treated using time-dependent perturbation theory, which results in transitions between the states established for the static interactions. For technological applications, the most important term is V_{EF} (see Eq. (1.38)), which gives rise to the emission and absorption of photons through radiative decay of the REIs. This includes the interaction between the electron charge and the electric field, as well as the interaction between the electron spin and the magnetic field.

The emission spectra of REIs indicate that the electromagnetic radiation is mainly from an electric dipole (ED), though in some cases magnetic dipole (MD) radiation may also be detected. Because the optical transitions take place between levels of a particular 4f^N configuration, ED radiation is forbidden in the first order, since the ED operator has uneven parity and the transition matrix element must have even parity (using Laporte selection rules). Van Vleck [52] pointed out that ED radiation can occur since the 4f^N states have admixtures of 4f^{N-1} nl configurations (where nl will be mostly 5d), and thus the 4f^{N-1} nl must be chosen such that it has opposite parity from 4f^N. These admixtures are produced via interactions that have odd parity and depend on the host material. Consequently, we can find four dominant sources of optical radiation from the REI spectra:

1. Forced ED induced by odd terms in the crystal field;
2. Forced ED radiation induced by the lattice's radiation;
3. Allowed MD transitions;
4. Allowed electric quadrupole radiation.

In free atoms, the MD radiation is ~10^6 orders of magnitude weaker than the ED radiation. Both types of radiation (ED and MD) depend on the host material of the REIs and can be detected from the emission spectra [53–58]. Thus, the quadrupole radiation is less probable in comparison with the MD radiation.

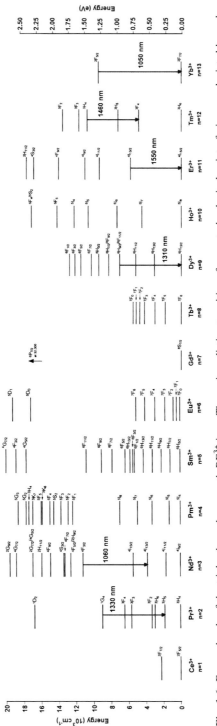

Fig. 1.6 Energy levels of the triply charged rare-earth RE^{3+} ions. The main radiative transitions from a technological point of view are depicted by *red arrows*. Figure adapted from Kenyon [51]

There are different types of transitions between levels (called ED, electric quadrupole, and MD), which are divided into allowed transitions (with a high probability) and forbidden transitions.

For the first-order allowed MD transition radiation, the free atom selection rules are still quite valid: $\Delta S = \Delta L = 0$ and $\Delta J = 0, \pm 1(0 \neq 0)$. For ED transitions, we have: $\Delta L = \pm 1, \Delta S = 0, |\Delta L|, |\Delta J| \leq 2l$. As ED transitions are induced by the crystal field, the selection rules break down almost completely, i.e., the selection rules with regard to the L, S, and J quantum numbers are now essentially governed by the crystal field interaction, yielding approximately $|\Delta J| \leq 6$. For the electric quadrupole case, the selection rules are: $\Delta S = 0, |\Delta L|, |\Delta J| \leq 2$.

The Judd–Ofelt theory is also outlined here because it has been intensively utilized for the parameterization of the intensities in REI crystal spectra since its publication more than 50 years ago [59, 60]. It is essentially a quantification of the ideas formulated by van Vleck [52] under optical radiation sources.

The total oscillator strength for a transition $i \rightarrow j$ is defined as:

$$f(i,j) = \frac{8\pi^2 m v}{3h(2J+1)} \left\{ \frac{(n_d + 2)^2}{9n_d} S_{ED} + n_d S_{MD} \right\}, \tag{1.42}$$

where S_{ED} and S_{MD} are the ED and MD intensities, respectively, defined as:

$$S_{ED}(\alpha J, \alpha' J') = \sum_{\lambda=2,4,6} \Omega_\lambda \left(\left\langle f^N \alpha[SL]J \left| U^{(\lambda)} \right| f^N \alpha'[S'L']J' \right\rangle \right)^2$$

$$S_{MD}(\alpha J, \alpha' J') = \beta^2 \sum_{\lambda=2,4,6} \Omega_\lambda \left(\left\langle f^N \alpha[SL]J \left| L + 2S \right| f^N \alpha' [S'L']J' \right\rangle \right)^2, \tag{1.43}$$

where $U^{(\lambda)}$ is a tensor operator of rank λ, and the sum runs over three terms of λ valued 2, 4, and 6. The Ω_λ parameters are expressed as follows:

$$\Omega_\lambda = (2\lambda + 1) \sum_{k(=1,3,5),q} \frac{\left| B_{k,q}^* \right|^2 |Y_{k,\lambda}|^2}{2k+1}. \tag{1.44}$$

So far, the Ω_λ parameters have been assumed to arise solely from the crystal field. However, they also contain contributions from admixtures with the lattice vibrations. In this manner, $B_{k,q}^*$ consist of odd-order parameters of the crystal field, radial integrals over wavefunctions of the $4f^n$ and perturbing, opposite parity wavefunctions of higher energy, and energies separating these states in terms of perturbation energy denominators. Here $Y_{k,\lambda}$ is given by:

$$Y_{k,\lambda} = 2\sum_{nl} \frac{\langle 4f|r|nl \rangle \langle nl|r^k|4f \rangle}{\Delta E_{nl}} \langle f\|C_{1,\lambda}\|l \rangle \langle l\|C_{k,\lambda}|f \rangle \left\{ \begin{matrix} 1 & k & \lambda \\ l & l & l' \end{matrix} \right\}. \tag{1.45}$$

In Eq. (1.45) for $Y_{k,\lambda}$, ΔE_{nl} is the energy difference between the 4f and opposite parity, nl configuration. The first two terms are the interconfiguration radial integrals: $\langle 4f|r|nl\rangle = \int_0^\infty R(4f)rR(nl)$ and $\langle nl|r^k|4f\rangle = \int_0^\infty R(nl)r^kR(4f)$. The second two terms are reduced matrix elements of the tensor operator $C_{k,q}$, which represents the angular part of the crystal field, and can be calculated from the angular momentum quantum numbers of the configurations according to the equations:

$$\langle f\|C_{1,\lambda}\|l\rangle = (-1)^l\sqrt{(2l+1)(2f+1)}\begin{Bmatrix} f & 1 & l \\ 0 & 0 & 0 \end{Bmatrix}$$

$$\langle l\|C_{k,\lambda}\|f\rangle = (-1)^f\sqrt{(2l+1)(2f+1)}\begin{Bmatrix} l & t & f \\ 0 & 0 & 0 \end{Bmatrix}$$

(1.46)

It is also of interest to determine the coefficient for spontaneous light emission from state $i(\alpha J)$ to state $j(\alpha' J')$. This is defined as:

$$A\left(\alpha J, \alpha' J'\right) = A_{ed} + A_{md} = \frac{64\pi^4 v^3}{3hc^2(2J+1)}\left\{\frac{n_d(n_d+2)^2}{9}S_{ED} + n_d^3 S_{MD}\right\}. \quad (1.47)$$

The Judd–Ofelt theory has been used to analyze a number of systems. Nevertheless, in most of these studies, the crystal field splitting terms were neglected. The total absorption intensities between the ground term and the excited terms are thus analyzed with only three empirical parameters $\Omega_\lambda (\lambda = 2, 4,$ and 6).

If j is an excited state that decays only by the emission of photons, then its observed relaxation rate is the sum of the probabilities for transitions to all possible final states. The $A(\alpha J, \alpha' J')$ is the reciprocal of the excited-state lifetime τ_j, as follows:

$$\tau_j = \frac{1}{A\left(\alpha J, \alpha' J'\right)}. \quad (1.48)$$

The branching ratio $\beta_{i,j}$ for the transition $j \to i$ is the fraction of all spontaneous decay processes that follow that channel, and it is defined as follows:

$$\beta_{i,j} = \frac{A(i,j)}{\sum_c A(i,j)} = A(i,j)\tau_j. \quad (1.49)$$

The branching ratio $\beta_{i,j}$ has an important effect on the performance of a device based on a specific transition, since it has a significant effect on the threshold of a laser and the efficiency of an amplifier.

1.5.2 Emission in the Near-Infrared Region

The optical region of the EM spectrum ranges from the deep ultraviolet (with vacuum ultraviolet starting at a wavelength of 10 nm) up to the far infrared, whose limit with the microwave domain is 1 mm. However, the far-infrared limit

Table 1.2 Current optical telecommunications bands

Band	Wavelength range (nm)	Description
O	1,260–1,360	Original wavelengths
E	1,360–1,460	Extended wavelengths
S	1,460–1,530	Short wavelengths
C	1,530–1,565	Conventional wavelengths
L	1,565–1,625	Long wavelengths
U	1,625–1,675	Ultra-long wavelengths

can be considered to be less than 100 μm. Indeed, there is currently a growing interest in THz-frequency technologies whose band is overlapping the far-infrared and microwave ranges (between 50 and 4,000 μm) [61]. Several spectral bands are then found within this optical region: the visible range (0.4–0.7 μm), the near-infrared range (NIR, 0.7–2.5 μm), the mid-infrared range (MIR, 2.5–10 μm), and the far-infrared range (FIR, 10–100 μm). Other spectral band classifications may also be found in the literature, depending on the application field. For instance, the wavelength bands used for optical telecommunications through optical fibers were standardized and are currently defined as in Table 1.2.

In optical telecommunications, multiple optical signals with distinct carrier frequencies are typically combined into a single-mode fiber to increase the transmission data rate. This technique is called wavelength division multiplexing (WDM), which can be categorized as dense WDM (DWDM) for use in long haul and ultra-long haul systems or as coarse WDM (CWDM) for use in metro systems, for example. The C- and L- bands, listed in Table 1.2, are the most important bands used today in optical telecommunications, particularly thanks to the ion emission window of erbium, as in Table 1.3. For further information, the reader is invited to see refs. [62, 63].

A large number of REIs may produce emissions in the optical telecommunications windows and, to a larger extent, in the near-infrared region from 0.7 to ~2.5 μm. Most of them may emit at different wavelengths according to the radiative transition involved. Table 1.3 lists the most commonly encountered REIs and their emission transitions and corresponding wavelengths. The most common host media for such REIs are yttrium aluminum garnet (YAG) single crystals, a wide variety of crystals based on lithium fluoride, and glasses based on silica, tellurite, or fluorides, which are largely exploited for their low-phonon properties if compared to oxide glasses.

Depending on the nature of the host (e.g., fluoride- or oxide-glass), some radiative transitions belonging to a single REI may or not be observed, as discussed in the details in ref. [64]. Another interesting property of the host that is that may strongly influence the spectroscopic properties of REIs is their nonlinear optical characteristics, more specifically their third-order nonlinear susceptibility, labeled $\chi^{(3)}$. For further reading about nonlinear optics, we must recommend the major reference work by Boyd [65]; we also recommend the more accessible book by New [66] and the one focused on nonlinear fiber optics by Agrawal [67]. It may

Table 1.3 Rare-earth ions, radiative transitions, and corresponding wavelengths for emission in the near-infrared region

Rare-earth ion	Transition	Wavelength (nm)
Erbium Er^{3+}	$^4S_{3/2} \rightarrow {}^4I_{13/2}$	~850
	$^4I_{11/12} \rightarrow {}^4I_{15/2}$	980–1,000
	$^4I_{13/2} \rightarrow {}^4I_{15/2}$	1,500–1,600
	$^2H_{11/2} \rightarrow {}^4I_{9/2}$	~1,660
	$^4S_{3/2} \rightarrow {}^4I_{9/2}$	~1,720
Thulium Tm^{3+}	$^3H_4 \rightarrow {}^3H_6$	803–825
	$^3H_4 \rightarrow {}^3F_4$	1,460–1,510
	$^1D_2 \rightarrow {}^1G_4$	~1,510
	$^3F_4 \rightarrow {}^3H_6$	1,700–2,015
	$^3H_4 \rightarrow {}^3H_5$	2,250–2,400
Praseodymium Pr^{3+}	$^3P_1 \rightarrow {}^1G_4$	880–886
	$^3P_1 \rightarrow {}^1G_4$	902–916
	$^1D_2 \rightarrow {}^3F_4$	1,060–1,110
	$^1G_4 \rightarrow {}^3H_5$	1,260–1,350
Holmium Ho^{3+}	$^5S_2, {}^5F_4 \rightarrow {}^5I_7$	~753
	$^5S_2, {}^5F_4 \rightarrow {}^5I_5$	~1,380
	$^5I_7 \rightarrow {}^5I_8$	2,040–2,080
Ytterbium Yb^{3+}	$^5F_{5/2} \rightarrow {}^5F_{7/2}$	970–1,040
Neodymium Nd^{3+}	$^4F_{3/2} \rightarrow {}^4I_{9/2}$	900–950
	$^4F_{3/2} \rightarrow {}^4I_{11/2}$	1,000–1,150
	$^4F_{3/2} \rightarrow {}^4I_{13/2}$	1,320–1,400

indeed be interesting to dope REIs in highly nonlinear optical media like tellurite glasses in order to exploit nonlinear processes such as two-photon or three-photon absorption and thus achieve efficient up-conversion emission frequencies.

Typically, fluorescence in REI-doped solids (if we consider a single-ion system) follows the well-known principle of Stokes' law, where excitation photons are at a higher energy than emitted ones, i.e., the excitation wavelength is lower than that of emission. In up-conversion emission, this principle is not followed: the energy of the emitted photons is higher than that of the excitation photons, or, in other words, the emission wavelength is lower than that of excitation. The involved mechanisms of this so-called anti-Stokes fluorescence have been well described by Auzel in [68]. Different nonlinear processes for excitation exist and lead to different types of up-conversion emission, for instance, cooperative luminescence, cooperative sensitization, the APTE (an acronym for the French name *Addition de Photon par Transfert d'Energie*) effect, second-harmonic generation (SHG), etc. [68]. Furthermore, energy transfer processes between identical ions or different ions can also lead to up-conversion fluorescence. For the latter, a well-known and widely studied example is the sensitizer role played by ytterbium ions, Yb^{3+}, in a medium containing other REIs like erbium, Er^{3+}, holmium, Ho^{3+}, and thulium, Tm^{3+}, which then acts as an actuator. Owing to the very large absorption cross section around 980 nm of the Yb^{3+} ions, efficient photon absorption can be achieved and is followed by energy transfers toward the resonant energy levels of the actuator ions, significantly increasing the quantum efficiency of specific radiative transitions.

Nowadays, access to laser diode sources around 980 nm at very competitive prices all around the world is allowing a great increase in research into the field of RE ions doped in luminescent materials. More specifically, sizable research efforts are devoted to finding new efficient sources for the near-infrared region, which is of primary interest for optical telecommunications technology based on silica optical fibers.

1.6 Perspective: Surface Plasmons in Nanophotonics

Since their conception by Kilby, Lehovec, Noyce, and Hoerni in the 1960s [69], integrated electronic circuits provide us with the ability to tailor, transport, and store electrons. However, the performance of such circuits is currently facing limitations when information must be carried at very high velocities, owing to their intrinsic losses [70, 71]. Photonics currently constitutes an efficient solution to this limit through the implementation of communication systems based on optical fibers and photonic circuits. Unfortunately, photonic components are still relatively large in volume (at the microscopic scale), limiting their integration into electronic chips (an optical integration at the nanoscale). There is thus a need to bridge the two technological areas of electronics and photonics for nanoscale applications.

Plasmonics thus appears to be an exciting and promising avenue of inquiry and it is explored by numerous research groups throughout the world, as shown in Fig. 1.7. Plasmonic circuits may join electronics and photonics at the nanoscale by providing an actual solution to the problems of nanoscale integration and data transfer rate. This new research area is so-called Nanophotonics and is dedicated to ultra-small optoelectronic components, offering extended bandwidths and high processing speeds (in the THz regime and the nanoscale). This technology may potentially revolutionize the telecommunications, computation, and sensing areas. For instance, IBM recently reported how to potentially exponentially increase the speed of processors, making current supercomputers faster than ever before [72]. This new technology, called CMOS (Complementary Metal Oxide Semiconductor) Integrated Silicon Nanophotonics, is based on the utilization of light pulses to accelerate data transmissions between chips, which can increase the speed of supercomputers about 1,000 times. Both electronic and optical modules are integrated on a single silicon wafer. Figure 1.7 shows the evolution of these three research areas.

Planar plasmonic devices are attracting interest for a myriad of applications owing to their potential compatibility with standard microelectronics technology and their ability to be densely integrated on a single chip. The challenges of using plasmonics in on-chip configurations reside in the required precise control over the plasmonic modes and their properties, particularly those related to their shape and size. Rare-earth-containing nanomaterials have recently shown great potential for

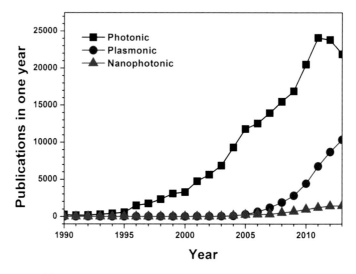

Fig. 1.7 A plot of the photonic, plasmonic, and nanophotonic publications per year. Data sourced from SCOPUS (April 2014)

applications in medicine, nanodevices, catalysis, fuel cells, etc., thanks to their unique optical and catalytic properties [4, 6, 21–23, 73, 74]. A large number of new efficient synthesis routes have been proposed and developed to produce various rare-earth-based nanomaterials of different nature, structure, size, and shape. Nano-structured rare-earth compounds can be obtained via well-defined techniques such as dry methods and aqueous and nonaqueous solution-based methods [75–77]. Considerable progresses has been made in producing such nanomaterials with high chemical purity, desired chemical composition and crystal phase, with controlled and uniform shape and size, with tuned surface status, and functionalization as well [75, 78, 79].

In this manner, nanophotonics is one of the most promising frontiers of knowledge, from not only a scientific but also a technological point of view, dealing with the interaction of light and matter at a lower scale than the light wavelength. Such interactions reveal in most of cases a new set of phenomena and revise the basic principles of optics, electrodynamics, quantum mechanics, optical microscopy, and linear and nonlinear interactions with matter, producing original challenges in fundamental research and new technologies.

References

1. Yehuda, B.: Light and matter: electromagnetism, optics, spectroscopy and lasers. Wiley, West Sussex (2006)
2. Menzel, R.: Photonics, linear and nonlinear interaction of laser light and matter, 2nd edn. Springer, Berlin (2007)

3. Novotny, L., Hecht, B.: Principles of Nano-Optics. Cambridge University Press, New York, NY (2006)
4. Quinten, M.: Optical Properties of Nanoparticles Systems. Wiley-VCH Verlag GmbH & Co. KGaA, Weinheim (2011)
5. Schmid, G. (ed.): Nanoparticles. Wiley-VSH Verlag GmbH & Co KGaA, Boschstr (2010)
6. Fendler, J.H. (ed.): Nanoparticles and Nanostructured Films. Wiley-VCH Verlag GmbH & Co KGaA, Weinheim (1998)
7. Gubin, S.P. (ed.): Magnetic Nanoparticles. Wiley-VCH Verlag GmbH & Co KGaA, Weinheim (2009)
8. Gaponenko, S.V.: Optical Properties of Semiconductors Nanocrystals. Cambridge University Press, Cambridge (1998)
9. Kreibig, U., Vollmer, M.: Optical Properties of Metal Clusters, Springer Series in Materials Science, vol. 25. Springer, Berlin (1995)
10. Raether, H.: Surface Plasmon on Smooth and Rough Surfaces and on Gratings, Springer Tracts in Modern Physics, vol. 111. Springer, New York, NY (1988)
11. Wood, R.W.: On a remarkable case of uneven distribution of light in a diffraction grating spectrum. Phil. Mag. **4**, 396 (1902)
12. Garnett, J.C.M.: Colours in metal glasses and in metallic films. Phil. Trans. Roy.Soc. Lond. **203**, 385–420 (1904)
13. Mie, G.: Beitrage zur optik truber medien. Ann. Phys. **25**, 377 (1908)
14. Pines, D.: Collective energy losses in solids. Rev. Mod. Phys. **28**, 184 (1956)
15. Fano, U.: Atomic theory of electromagnetic interactions in dense materials. Phys. Rev. **103**, 1202 (1956)
16. Ritchie, R.H., Arakawa, E.T., Cowan, J.J., Hamm, R.N.: Surface-Plasmon resonance effect in grating diffraction. Phys. Rev. Lett. **21**, 1530 (1968)
17. Kretschmann, E., Raether, H.: Radiative decay of non radiative surface plasmons excited by light. Z. Naturforsch. A. **23**, 2135 (1968)
18. Otto, A.: Excitation of nonradiative surface plasma waves in silver by the method of frustrated total reflection. Z. Physik. **216**, 398 (1968)
19. Maradudi Cunningham, C., Wallis, R.F.: Effect of a charge layer on the surface-plasmon-polariton dispersion curve. Phys. Rev. B **10**, 3342 (1974)
20. Maier, S.A.: Plasmonics: Fundamentals and Applications. Springer Science + Business Media LLC, New York, NY (2007)
21. Brongersma, M.L., Kik, P.G.: Surface Plasmon Nanophotonics. Springer, Dordrecht (2007)
22. Gaponenko, S.V.: Introduction to Nanophotonics. Cambridge University Press, Cambridge (2010)
23. Ohtsu, M. (ed.): Progress in Nanophotonics. Springer, New York, NY (2013)
24. Jahr, N., Anwar, M., Stranik, O., Ha¨drich, N., Vogler, N., Csaki, A., Popp, J., Fritzsche, W.: Spectroscopy on single metallic nanoparticles using subwavelength apertures. J. Phys. Chem. C **117**(15), 7751 (2013)
25. Lassiter, J.B., et al.: Plasmonic waveguide modes of film-coupled metallic nanocubes. Nano Lett. **13**(12), 5866 (2013)
26. Wen, J., Romanov, S., Peschel, U.: Excitation of plasmonic gap waveguides by nanoantennas. Opt. Exp. **17**(8), 5925 (2009)
27. Bozhevolnyi, S.I., Sondergaard, T.: General properties of slow-plasmon resonant nanostructures: nano-antennas and resonators. Opt. Exp. **15**(17), 10869 (2007)
28. Anderson, L.J.E., et al.: A tunable plasmon resonance in gold nanobelts. Nano Lett. **11**(11), 5034 (2011)
29. Jensen, T.R., et al.: Nanosphere lithography: Tunable localized surface plasmon resonance spectra of silver nanoparticles. J. Phys. Chem. B **104**(45), 10549 (2000)
30. Novoty, L., Hecht, B.: Principles of Nano-Optics, 2nd edn. Cambridge University Press, New York, NY (2006)

31. Lindquist, N.C., et al.: Nanoantennas for visible and infrared radiation. Rep. Prog. Phys. **75**(3), 024402 (2012)
32. Ordal, M.A., Long, L.L., Bell, R.J., Bell, S.E., Bell, R.R., Alexander Jr., R.W., Ward, C.A.: Optical properties of the metals Al, Co, Cu, Au, Fe, Pb, Ni, Pd, Pt, Ag, Ti, and W in the infrared and far infrared. Appl. Opt. **22**(7), 1099 (1983)
33. Palik, E.D. (ed.): Handbook of Optical Constants of Solids II. Elsevier, Orlando (1998)
34. Serna, R., Gonzalo, J., Afonso, C.N., de Sande, J.C.G.: Anomalous dispersion in nanocomposite films at the surface plasmon resonance. Appl. Phys. B. **73**(4), 339 (2001)
35. Gonzalez-Tudela, A., Huidobro, P.A., Martín-Moreno, L., Tejedor, C., García-Vidal, F.J.: Reversible dynamics of single quantum emitters near metal-dielectric interfaces. Phys. Rev. B **89**, 041402(R) (2014)
36. Ford, G., Weber, W.: Electromagnetic-interactions of molecules with metal-surfaces. Phys. Rev. **113**, 195 (1984)
37. Barnes, W.: Fluorescence near interfaces: The role of photonic mode density. J. Mod. Opt. **45**(4), 661 (1998)
38. Rivera, V.A.G., Osorio, S.P.A., Ledemi, Y., Manzani, D., Messaddeq, Y., Nunes, L.A.O., Marega Jr., E.: Localized surface plasmon resonance interaction with Er^{3+}-doped telluriteglass. Opt. Exp. **18**(24), 25321 (2010)
39. Guo, Y., Zhang, Y.-F., Bao, X.-Y., Han, T.-Z., Tang, Z., Zhang, L.-X., Yuan, Z., Gao, S.: Linear-response study of plasmon excitation in metallic thin films: layer-dependent hybridization and dispersion. Phys. Rev. B **73**, 155411 (2006)
40. Zhang, Y.-F., Jia, J.-F., Han, T.-Z., Tang, Z., Shen, Q.-T., Guo, Y., Qiu, Z.Q., Xue, Q.-K.: Band structure and oscillatory electron-phonon coupling of Pb thin films determined by atomic-layer-resolved quantum-well states. Phys. Rev. Lett. **95**, 096802 (2005)
41. Nielsen, M.A., Chuang, I.L.: Quantum Computation and Quantum Information. Cambridge University Press, Cambridge (2000)
42. Empedocles, S.A., Neuhauser, R., Bawendi, M.G.: Three-dimensional orientation measurements of symmetric single chromophores using polarization microscopy. Nature **399**, 126 (1999)
43. Basche, T., Moerner, W., Orrit, M., Wild, U.: Single–Molecule Optical Detection, Imaging and Spectroscopy. VCH, Verlagsgesellschaft, Weinheim (1977)
44. Thyssen, P., Binnemans, K.: Handbook on the Physics and Chemistry of Rare Earths, Chapter 248, vol. 41. Elsevier, AE Amsterdam, The Netherlands (2011)
45. Liu, G., Jacquier, B. (eds.): Spectroscopic Properties of Rare-Earths in Optical Materials, vol. 83. Series: Springer Series in Materials Science, Berlin (2005)
46. Hufner, S.: Optical Spectra of Transparent Rare Earth Compounds. Academic, New York, NY (1978)
47. Judd, B.R.: Operator Techniques in Atomic Spectroscopy. McGraw-Hill, New York, NY (1963)
48. Wybourne, B.G.: Spectroscopic Properties of Rare Earths. Wiley, New York, NY (1965)
49. Pisarski, W.A., Goryczka, T., Wodecka-Dus, B., Płonska, M., Pisarska, J.: Structure and properties of rare earth-doped lead borate glasses. Mat. Sci. Eng. B. **122**(2), 94 (2005)
50. Kanjilal, A., Rebohle, L., Skorupa, W., Helm, M.: Correlation between the microstructure and electroluminescence properties of Er-doped metal-oxide semiconductor structures. Appl. Phys. Lett. **94**, 101916 (2009)
51. Kenyon, A.J.: Recent developments in rare-earth doped materials for optoelectronics. Prog. Quantum. Electron. **26**, 225 (2002)
52. van Vleck, J.H.: The puzzle of rare-Earth spectra in solids. J. Phys. Chem. **41**, 67 (1937)
53. Cheng, Z., Xing, R., Hou, Z., Huang, S., Jun, L.: Patterning of light-emitting YVO_4:Eu^{3+} thin films via inkjet printing. J. Phys. Chem. C **114**(21), 9883 (2010)
54. Rivera, V.A.G., Ferri, F.A., Clabel, H.J.L., Pereira-da-Silva, M.A., Nunes, L.A.O., Li, M.S., Marega Jr., E.: High red emission intensity of Eu:Y_2O_3 films grown on Si(100)/Si (111) by electron beam evaporation. J. Lumin. **148**, 186 (2014)

55. Tanabe, S., Ohyagi, T., Todoroki, S., Hanada, T., Soga, N.: Relation between the Ω_6 intensity parameter of Er^{3+} ions and the Eu isomer shift in oxide glasses. J. Appl. Phys. **73**, 8451 (1993)

56. Tanabe, S.: Optical transitions of rare earth ions for amplifiers: how the local structure works in glass. J. Non-Cryst. Solids **259**(1–3), 1 (1999)

57. Lin, H., Yang, D., Liu, G., Ma, T., Zhai, B., An, Q., Yu, J., Wang, X., Liu, X., Pun, E.Y.B.: Optical absorption and photoluminescence in Sm^{3+}- and Eu^{3+}-doped rare-earth borate glasses. J. Lumin. **113**(1–2), 121 (2005)

58. Luciana, R.P.K., da Silva, D.S., Luciana, C.B., de Araújo, C.B.: Influence of metallic nanoparticles on electric-dipole and magnetic-dipole transitions of Eu^{3+} doped germinate glasses. J. Appl. Phys. **107**, 113506 (2010)

59. Judd, B.R.: Optical absorption intensities of rare-earth ions. Phys. Rev. **127**, 750 (1962)

60. Ofelt, G.S.: Intensities of crystal spectra of rare-earth ions. J. Chem. Phys. **37**, 511 (1962)

61. Lee, Y.-S.: Principles of Terahertz Science and Technology. Springer, New York, NY (2009)

62. Agrawal, G.P.: Fiber-Optic Communication Systems. Wiley, New York, NY (2002)

63. Mitschke, F.: Fiber Optics, Physics and Technology. Springer, New York (2009)

64. Digonnet, M.: Rare-Earth Doped Fiber Lasers and Amplifiers, Stanford, Chapter 2, 2nd edn. Marcel Dekker, New York, NY (2001)

65. Boyd, R.W.: Nonlinear Optics, 3rd edn. Elsevier, London, UK (2008)

66. New, G.: Introduction to Nonlinear Optics. Cambridge University Press, Cambridge (2008)

67. Agrawal, G.: Nonlinear Fiber Optics, 4th edn. Academic, New York, NY (2006)

68. Liu, G., Jacquier, B.: Spectroscopic Properties of Rare-Earths in Optical Materials, Chapter 5. Springer, New York, NY (2005)

69. Kilby, J.S.: Invention of the integrated circuit. IEEE Trans Electron Dev **23**, 648 (1976)

70. Agrawal, G.P.: Fiber-Optic Communication Systems, 3rd edn. Wiley, New Jersey (2010)

71. Mitschke, F.: Fiber Optics, Physics and Technology. Springer, New York, NY (2009)

72. Vlasov, Y.A.: IEEE. Comm. Mag. **50**(2), S67–S72 (2012)

73. Tanabe, K.: A review of ultrahigh efficiency III-V semiconductor compound solar cells: Multijunction tandem, lower dimensional, photonic up/down conversion and plasmonic nanometallic structures. Energies **2**(3), 695 (2009)

74. Rivera, V.A.G., Ferri, F.A., Marega, E. Jr.: In: Kim KY (ed.) Localized Surface Plasmon Resonances: Noble Metal Nanoparticle Interaction with Rare-Earth Ions, Chapter 11, Intech, Croatia (2012)

75. Functionalized Inorganic Fluorides, Synthesis, Characterization and Properties of Nanostructured Solids. In: Tressaud A. (ed.). Wiley, United Kingdom (2010)

76. Yan, C.-H., et al.: Handbook on the Physics and Chemistry of Rare Earths, (Chapter 251), vol. 41. Elsevier, New York, NY (2011)

77. Wang, F., Liu, X.: Rare-earth doped upconversion nanophosphors. In: Andrews, D., Scholes, G., Wiederrecht, G. (eds.) Comprehensive Nanoscience and Technology, (Chapter 18), vol. 1. Elsevier, New York, NY (2010)

78. Wang, F., Liu, X.: Recent advances in the chemistry of lanthanide-doped upconversion nanocrystals. Chem. Soc. Rev. **38**, 976 (2009)

79. Wang, F., Han, Y., Lim, C.S., Lu, Y.H., Wang, J., Xu, J., Chen, H.Y., Zhang, C., Hong, M.H., Liu, X.G.: Simultaneous phase and size control of upconversion nanocrystals through lanthanide doping. Nature **463**, 1061 (2010)

Chapter 2
Plasmonic Nanoparticles Coupled with an |n⟩-State Quantum System

2.1 Introduction

The first chapter was dedicated to the interaction of light with matter via the description of the dielectric function of a free electron gas, the optical properties of metals and rare-earth ions (REI), and the surface plasmon polariton (SPP) in a gain medium, thus laying the foundation for a good understanding of the field and easier reading of this book. In this chapter, we will describe the behaviors resulting from coupling between plasmonic nanoparticles and an |n⟩-state quantum system. By definition, an n-state quantum system is a system characterized by a set of quantum numbers, represented by an eigenfunction, and for which the energy of each state is precisely within the limits imposed by the uncertainty principle but may be changed by applying a field or force. States of the same energy are called degenerate [1]. Solid-state emitters, such as semiconductor quantum dots (QD) or REIs, are examples of n-state quantum systems that have been extensively investigated. In particular, REIs have been recently attracting much interest for quantum information processing, owing to their unique shielding of 4f-shell transitions from their surroundings and consequently longer coherence times [2, 3].

The coupling between nanoparticles (NPs) of noble metals with an n-state quantum system like an REI may influence the optical properties of the latter and particularly enhance their emission features, opening new avenues and possibilities for practical applications (e.g., in telecommunications [4, 5]). The mechanism for the interactions between NPs and REIs that are responsible for this enhanced emission is however still open to debate: sometimes it is ascribed to energy transfer processes between them, and sometimes it is attributed to an improvement of the local field around the NPs. In fact, both effects are produced from a coherent collective oscillation of electrons at the surface of the metal NP, also known as Localized Surface Plasmon Resonance (LSPR) [6].

In the first section of this chapter, we will describe plasmonic NPs and their utilization as optical nanoantennas by defining the plasmon modes, resonance

© V.A.G. Rivera, O.B. Silva, Y. Ledemi, Y. Messaddeq, and E. Marega Jr. 2015
V.A.G. Rivera et al., *Collective Plasmon-Modes in Gain Media*, SpringerBriefs
in Physics, DOI 10.1007/978-3-319-09525-7_2

frequency, and some specific properties of the NPs. Recent advances related to this field can be found in special journal issues, in refs. [7, 8]. Next, we will discuss in detail the interaction between plasmonic NPs and quantum emitters (QE) (with special interest in REIs).

2.2 Plasmonic Nanoparticles: Optical Nanoantennas

As described in the previous chapter, plasmonic NPs, i.e., NPs of noble metals supporting plasmons, possess unique properties and can provide their host material with specific and extraordinary optical properties. In this section, emphasis is given to these enhanced optical properties provided to a host medium, such as optical gain, i.e., allowing optical emission and/or amplification via REI doping, for instance. Surface plasmon resonances and their effects, like local electrical field changes, are then exploited to tailor and enhance the spectroscopic properties of the optically active entities (the REIs in this case), leading to new material properties (as multifunctional materials). Plenty of fundamental and applied research is performed all over the world in this field, as we will see throughout this chapter.

A new way to utilize plasmonic NPs has emerged over the last few years: exploiting the various shapes and dimensions that can be given to them to create optical (nano-) antennas. By definition, optical antennas are analogous to the radiofrequency or microwave antennas, but their working range is in the visible or near-infrared (optical) domain, i.e., at very high frequencies (hundreds of terahertz). An optical antenna can be defined as an antenna that converts efficiently localized energy (from the near field) into propagating radiation (to the far field), and vice versa [9]. An optical antenna is a device or object through which a receiver and a transmitter interact via free optical radiation. The general principle of an optical antenna is illustrated in Fig. 2.1. Here, the receiver or transmitter is an elemental quantum absorber or emitter like a noble metal NP, but it can also be an atom, molecule, REI, QD, or a defect center.

The primary function of an antenna is to boost the interaction between an emitter/absorber and the radiation field. Such a function can therefore be easily applied to a quantum system by controlling the light–matter interaction and more particularly to metal nanostructures like NPs, which behave as plasmons at optical frequencies. In this manner, a plasmonic nanostructure can convert local frequencies with high efficiency, making it possible to exploit frequency-selective interaction across the visible and infrared spectrum while requiring excitation of only a single frequency.

This conversion-field ability can be used to interact with QEs at the nanoscale. The challenge for plasmonic nanostructure design is that the structures need to resonate simultaneously at multiple wavelengths, i.e., the wavelengths of incoming and outgoing radiation.

This emerging field is attracting a lot of interest in improving the performance and efficiency of devices in a number of applications, such as light emission, photovoltaics, spectroscopy, photodetection, and sensing. Comprehensive reviews on this field have been published by Novotny et al. [9].

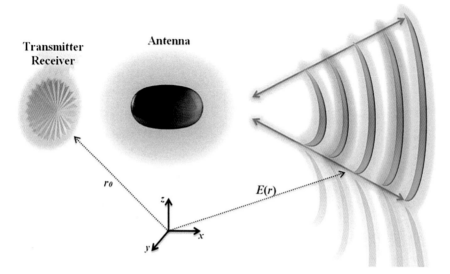

Fig. 2.1 The general principle of optical antenna theory. A transmitter or receiver (atom, ion, molecule, etc.) interacts with free optical radiation via an optical antenna

2.2.1 Plasmon Modes and Resonance Frequencies

The strong coupling between surface charges, resulting from the oscillations of electrons and the EM wave at the planar metal–dielectric interface, constitutes a propagating wave called an SPP, as shown in Fig. 1.3a. Nevertheless, in this section, we are interested in understanding the interaction of plasmons in NPs.

Applications in cancer therapy [10, 11] and molecular sensing [12, 13], for instance, are based on effects derived from interactions between NPs and plasmons. In contrast to the SPP mode, in which the excitation of free electrons by EM radiation arises from the propagation of a wave along the metal–dielectric interface, plasmon–NP interaction is characterized by a local excitation in the surface of the NPs. Therefore, this plasmon mode is known as a localized surface plasmon (LSP). The formation of such a mode in the vicinity of NPs, as the SPP mode, is described theoretically by Maxwell's equations. However, before we start to look for solutions, it is necessary to consider the effects resulting from the geometry of the NPs, because this factor establishes the main features of how NPs interact with the EM fields.

Initially, in order to avoid changes in the phase of the driving field, we will consider particles with dimensions d, where $d \ll \lambda$, and λ is the wavelength of the incident radiation. Such a condition is required for simplicity in the analytical solution for the fields; since there is no variation in the phase, it is possible to regard the case of an NP in an electrostatic field. Furthermore, for simplicity, we consider the example of a metallic nanosphere [14] with a permittivity ε_m and radius R embedded within a dielectric medium of permittivity ε_d. To obtain a consistent solution for the field it is necessary to solve the Laplace equation $\nabla^2 \phi = 0$. Since the nanosphere presents azimuthal symmetry, the electric potential is a function of distance r from the origin and of the polar angle θ. Thus, the general solution for the Laplace equation is written as follows:

$$\phi(r, \theta) = \sum_{i=0}^{\infty} \left(C_l r^l + D_l r^{-(l+1)} \right) P_l(\cos \theta), \tag{2.1}$$

where the coefficients C_l and D_l are determined by the boundary conditions and the terms $P_l(\cos \theta)$ are the Legendre polynomials. It is reasonable to consider the electric field as uniform at distances far away from the nanosphere (as $r \to \infty$), i.e., $\boldsymbol{E} = E_0 \hat{z}$. From the relations between the electric field and potential ($\boldsymbol{E} = -\nabla \phi$ or $\phi = -\int \boldsymbol{E} \cdot d\boldsymbol{r}$), for great distances from the nanosphere, the electric potential is defined as $\phi = -E_0 z = -E_0 r \cos \theta$. This approximation imposes certain conditions for the coefficient C_l:

$$\begin{cases} C_l = 1 \text{ for } l = 1 \\ C_l = 0 \text{ for } l \neq 1 \end{cases} . \tag{2.2}$$

The necessary approach to determine the coefficient D_l arises from the boundary conditions at the interface of the nanosphere for the parallel components of the electric field, and the normal component of electric displacement. From Eq. (2.1), we notice that for regions where $r < R$ (inside the sphere) $D_l = 0$ for any l, otherwise the potential would diverge. Furthermore, since the nanosphere is metallic, the electric field vanishes inside the conductor. Thus, we are interested in solutions outside the nanosphere. After the application of boundary conditions, we obtain the following electric potential:

$$\phi = -E_0 r \cos \theta + \frac{\varepsilon_m - \varepsilon_d}{\varepsilon_m + 2\varepsilon_d} E_0 R^3 \frac{\cos \theta}{r^2}. \tag{2.3}$$

We can rewrite this expression in terms of dipole momentum $\boldsymbol{p}$, as:

$$\phi = -E_0 r \cos \theta + \frac{\boldsymbol{p} \cdot \boldsymbol{r}}{4\pi \varepsilon_0 \varepsilon_d r^3}. \tag{2.4}$$

Comparing the last two equations, the dipole momentum is written as:

$$\boldsymbol{p} = 4\pi \varepsilon_0 \varepsilon_d R^3 \frac{\varepsilon_m - \varepsilon_d}{\varepsilon_m + 2\varepsilon_d} \boldsymbol{E}. \tag{2.5}$$

From Eq. (2.4), it is important to note that $\boldsymbol{p}$ is induced on the nanosphere resulting from the application of an external electric field, and the dipole's strength depends on the dispersive properties of the materials. The polarizability α is defined from the ratio between $\boldsymbol{p}$ and the electric displacement $\boldsymbol{D} = \varepsilon_0\varepsilon_d\boldsymbol{E}$:

$$\alpha = 4\pi R^3 \frac{|\varepsilon_m - \varepsilon_d|}{|\varepsilon_m + 2\varepsilon_d|}. \tag{2.6}$$

The polarizability α expresses the capacity of the nanosphere to couple the plasmon mode on its surface with the dipole momentum $\boldsymbol{p}$ induced by the external electric field. As we discussed in the beginning of this section, the excitation of the charge density has local effects, like polarizability, since it depends on the radius of the nanosphere (R). A maximum value for α is achieved when the absolute value of the denominator in the right-hand side of Eq. (2.6) is at a minimum. This condition defines the resonance regime for metallic nanospheres embedded in a dielectric medium:

$$\mathrm{Re}[\varepsilon_m] = -2\varepsilon_d. \tag{2.7}$$

At frequencies for which this result is valid, there is an enhancement of the plasmon mode along the surface of the nanosphere. Thus, this mode is defined as localized surface plasmon resonance, or LSPR. The surface charge of the NP oscillates along the direction of polarization of the incident electric field at resonance. The behavior of this plasmon mode is similar to the case of an electric field due to a dipole momentum.

In particular, when an REI is added into a host material there is an increase in the permittivity [15], $n_d^2 = \varepsilon_d(\omega)$. Consequently, the sum of the real part in the denominator from Eq. (2.6) is minimized, since the real part of the permittivity for metals is a negative quantity (see Figs. 1.1 and 1.2). Thus, in this new environment, the polarizability α (see Eq. (2.6)) of the NP increases at resonance, including when we compare the resonance of the case for a regular dielectric medium, i.e., $\mathrm{Re}[\varepsilon_d] > 0$.

The present results are valid for the specific case of a nanosphere smaller than the wavelength of the incident electric field. Other situations, in which the dimensions of the NP are comparable to the wavelength and present different geometries, will be discussed in Sect. 2.3.3 using Mie theory for light scattering. Furthermore, the plasmon–photon interaction for a general metallic NP will be developed in Sect. 2.4, where we will use the Green's function formalism.

2.3 Analysis of Some Specific Properties of Nanoparticles

In order to understand and manipulate SPPs and LSPRs, we must be able to fully characterize and model the plasmonic properties of metallic nanostructures. To achieve this goal, a number of approaches have been developed to detect and analyze both the far- and near-field properties of plasmonic nanostructures. Parts of these approaches are based on traditional optical techniques, such as dark-field microscopy and UV–Vis–NIR spectroscopy (in the far-field). Other parts use sophisticated new technologies to obtain extremely detailed information; for instance, scanning near-field optical microscopy (SNOM), which offers spatial resolutions beyond the diffraction limit (in the near-field), making it ideal for imaging plasmonic nanostructures. Furthermore, topographic images can be collected simultaneously using traditional atomic force microscopy (AFM) imaging. Furthermore, tips used for AFM or for scanning tunneling microscopy (STM) can be used as probes.

Computational modeling and simulations are also powerful tools for understanding the properties of plasmonic nanostructures. The fundamental way to study the plasmonic properties of a plasmonic nanostructure is to solve Maxwell's equations for the interaction of light and matter at the nanoscale. For example, for a homogeneous nanosphere and an incident plane wave of light, Mie theory gives a simple, analytical solution for the space inside and outside of the spherical boundary. Nevertheless, for more complicated shapes, an exact solution cannot be reached owing to the reduced symmetry. Therefore, numerical methods have been established for determining the plasmonic properties of these metallic nanostructures. Well-known examples include the discrete dipole approximation (DDA) [16], the finite-difference time domain (FDTD) method [17], and the finite element method (FEM) [18]. These numerical methods can also be used to plot the near-field distributions of electric fields in space over the volume of the nanostructure.

As discussed above, there are many theoretical and experimental techniques available to study metallic nanostructures and their properties. Currently, there are efforts to extend the capacity of these methods in order to further exploit the abilities of new nanodevices based on plasmonics with important applications. Below we will discuss three key points of the optical properties of metallic NPs.

2.3.1 Effect of the Dielectric Environment

Firstly, it is worth describing the effect of the medium in which the metal NPs are embedded and, in particular, its dielectric properties. The host medium must be a dielectric, i.e., an electrical insulator that can be polarized under an electric field, to make possible the existence of LSPs on metal NPs. The dielectric properties of the host medium, particularly its refractive index close to the particle surface, will then govern both the frequencies and intensities of the LSPs. The band wavelength and refractive index of the LSPs are linked through the relation:

$$n_{\mathrm{d}} = \sqrt{\varepsilon_{\mathrm{d}} \cdot \mu_{\mathrm{d}}}, \tag{2.8}$$

where μ_d is its relative permeability. The permeability is a physical quantity accounting for the ability of a material to support a magnetic field. In dielectric media, such as glass, water, gas, etc., the permeability μ_d is equal to 1, giving thus

$n_d = \sqrt{\varepsilon_d}$. Next, ε_R is the material's relative permittivity, and is equal to the ratio of the material's absolute permittivity ε to the vacuum permittivity ε_0 ($\varepsilon_0 = 8.8542 \times 10^{-12}$ F/m):

$$\varepsilon_R = \frac{\varepsilon_d}{\varepsilon_0}. \tag{2.9}$$

From Eq. (1.9) in Sect. 1.2, which gives the dielectric function $\varepsilon(\omega)$ as a function of frequency from the Drude model for the electronic structure of metals, a simplified relation can be proposed by considering that $\gamma \ll \omega_p$ for visible and infrared frequencies:

$$\varepsilon(\omega) = 1 - \frac{\omega_p^2}{\omega^2}. \tag{2.10}$$

At resonance (with Frohlich conditions, Eq. (2.7)), we have:

$$\omega_{max} = \frac{\omega_p}{\sqrt{2\varepsilon_m + 1}}, \tag{2.11}$$

where ω_{max} is the frequency of the LSPR peak. If we convert from frequency to wavelength via $\lambda = 2\pi c/\omega$, then the above expression becomes:

$$\lambda_{max} = \lambda_{spp} \sqrt{2n_m^2 + 1}, \tag{2.12}$$

where λ_{max} is the LSPR peak wavelength and λ_{spp} is the wavelength corresponding to the plasma frequency of the bulk metal. The dependence of the LSPR peak wavelength on the refractive index of the surrounding medium tends to be linear at optical frequencies over small ranges of n_d [19].

Further investigations about effects of the dielectric environment on the properties of metal nanoparticles are reported in refs. [19–22].

2.3.2 Effect of the Composition

As we have seen in previous sections, the effects of SPPs present a great number of applications, like the enhancement of the wave guiding modes in metallic waveguides [23] and optical transmission through sub-wavelength structures [24–26], for instance. Such effects constitute a coupling of the electric charge density on the metal's surface with the EM field. Additionally, other applications, like the development of novel materials for solar cells that can replace the well-known cells based on silicon for improvement of radiation absorption [27], and techniques for molecule sensing [28] and even cancer diagnostics [29], for instance, are no longer explained by the conventional SPP model. Thus, to comprehend effects arising

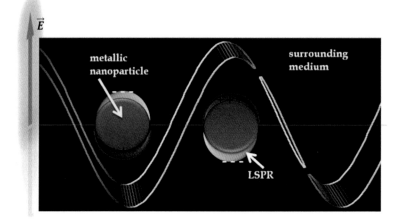

Fig. 2.2 The electron charge density of an NP is displaced by an external electric field

from these applications, it is necessary to understand another type of light–matter interaction in a plasmonic field: LSPR.

LSPR results from the coupling of electrons present on the surface of NPs with EM radiation (see Fig. 2.2). The confinement of plasmons in regions with dimensions comparable to or less than the wavelength enhances the local electric field at the nanoscale. The intensity of the electric field associated with LSPR decays rapidly with distance from the center of the NP. An example of this fact is depicted in ref. [30].

Noble metal NPs exhibit plasmon resonance frequencies in the visible region of the spectrum, which coincide with the maximum values of their optical extinction. The wavelength related to this maximum depends on the surrounding medium in which the NPs are embedded. In this framework, sensing measurements based on LSPR are widely used. Commonly used material for studies and applications of LSPR are Au and Ag NPs, due to their particular properties. Gold exhibits great chemical stability and is also resistant to oxidation effects. Silver has sharp resonance curves, i.e., it is a low-loss medium compared to other noble metals, and furthermore, presents a high refractive index sensitivity, which provides excellent conditions for the development of sensors. Although these noble metals display such features, they also have losses due to their conductivity and the imaginary parts of their dielectric functions (see Chap. 1, Sect. 1.2), which is a critical condition for the development of optical devices. To overcome this issue, LSPR effects have been measured using alternative materials for the NPs [31, 32].

Thus far, the focus of this section has been almost entirely on Au and Ag as plasmonic NPs, owing to their distinct dielectric properties in the Vis–NIR spectral range. In particular, the low intrinsic losses through intraband excitations at energies just below the interband absorption threshold, which are related to the position of the d bands with respect to the Fermi level, generally make Ag and Au very attractive plasmonic metals. However, other metals also deserve similar attention as "novel" plasmonic materials, owing to their different bulk dielectric properties, compared to Ag and Au, with a strong interband activity over a wide frequency range and concomitant damping. For instance, the interband activity varies greatly between Au, Pt, and Al. Au has a threshold for interband excitation at ~540 nm. For Pt, interband transitions are an active channel for LSPR decay at all energies, because the top of the 4d band overlaps with the Fermi level. In Al, interband excitations are present only within a narrow energy range (around 826 nm owing to a pair of parallel bands around the Σ axis on the Γ–K–W–X plane of the Brillouin zone). Furthermore, the electron density of the conduction band of Al is three times greater than that of Au, resulting in a shift to higher LSPR energies [33]. For more details about the plasmonic properties of Al and, in particular, the LSPR spectroscopy of Al NPs, see refs. [34, 35].

Chan et al. [36] presented a comparison of Al, Ag, Cu, and Au triangular NPs for a similar shape and geometry, showing that the LSPR λ_{max} has the ordering Au > Cu > Ag > Al, while the full width at half maximum (damping) satisfies Al > Au > Ag > Cu.

In order to improve the characteristics of nanoscale materials as well as to expand the light scattering and absorption properties of Au or Ag NPs, new nanostructures can be achieved by incorporating a second metal. These advances are gaining increasing interest in the field of multifunctional plasmonics, where it is now possible to correlate nanocrystal structures with optical properties such as LSPR. Thus, composition provides a synthetic lever for manipulating multifunctional plasmonic properties. More details about multifunctional plasmonics can be found in the work of DeSantis et al. [37] and in Chemical Reviews [38].

2.3.3   Effect of the Shape and Size

Being able to control the properties of plasmonic NPs means being able to manipulate its dimensions or shape. The size and shape of an NP determine its plasmonic features, including the ratio of absorption to scattering, the number of LSPR modes (horizontal or transversal), and the peak position of an LSPR mode [6, 39–41]. Therefore, size and shape are important parameters that must be carefully considered to achieve a specific application.

Thus far, we have discussed the resonance modes on flat metallic surfaces (Chap. 1). Nonetheless, in this issue, it is also necessary to speak about bounded surfaces with curved geometries (e.g., NPs), which demonstrate an effective restoring force on the driven electrons (in LSPR) that give rise to resonance on the surface. Moreover, the parameters of metal nanostructures (e.g., composition, size, shape, and environment) have an influence on the ω_{spp}. In metals such as Pb, In, Hg, Sn, and Cd, the resonance frequency lies in the UV spectral region, and small metal structures do not display a strong color effect. Likewise, such small metal structures are also readily oxidized, which strongly prevents LSPR excitation. However, the noble metals (e.g., Ag and Au) are exceptional because of their good stability in air, and their LSPR frequency is pushed into the visible parts of the spectrum owing to their d–d band transitions. Hence, the excitation of LSPR is most commonly performed in noble metal nanostructures; thus, current reports on metal nanostructures for LSPR mainly cover noble metal nanostructures. The configurations of these nanostructures for LSPR involve mainly NPs, nanorods, nanowires, nanosheets, and nanodisks.

From Mie theory, we can write the extinction cross section of a spherical NP as:

$$\sigma_{ext} = \frac{\lambda^2}{2\pi} \sum_{n=0}^{\infty} (2n + 1) \mathrm{Re}\{a_n + b_n\}. \tag{2.13}$$

The parameters a_n and b_n are defined as:

$$a_n = \frac{\Psi_n(\beta)\Psi'_n(m\beta) - m\Psi_n(m\beta)\Psi'_n(\beta)}{\xi_n(\beta)\Psi'_n(m\beta) - m\Psi_n(m\beta)\xi'_n(\beta)} \tag{2.14}$$

and

$$b_n = \frac{m\Psi_n(\beta)\Psi'_n(m\beta) - \Psi_n(m\beta)\Psi'_n(\beta)}{m\xi_n(\beta)\Psi'_n(m\beta) - \Psi_n(m\beta)\xi'_n(\beta)} \tag{2.15}$$

where $\beta\left(= \frac{\pi d n_d}{\lambda}\right)$ is the size parameter, the Riccati–Bessel functions Ψ_n and ξ_n are the half-integer-order Bessel functions of the first kind $(J_{n+1/2}(z))$, $\Psi_n(x) = \sqrt{\frac{\pi x}{2}}J_{n+0.5}(x)$, $\xi_n(x) = \sqrt{\frac{\pi x}{2}}H_{n+0.5}(x)$, and $H_{n+0.5}(x)$ is the half-integer-order Hankel function of the second kind.

In order to apply this concept, we will focus our attention on the silver and gold metals, since the LSPR condition mentioned above is satisfied at visible light frequencies, as in Fig. 2.3. The dielectric constants for Ag and Au were extracted from Palik [42], and the medium dielectric constant is assumed to be 1 (i.e., a particle in a vacuum medium) and 2 for NPs with different sizes. Figure 2.3 shows the dependence of ω_p on size of the NPs (Ag and Au) and the refraction index of the host material. Using Mie theory is not a very computationally expensive calculation, and the quasistatic expressions are suitable to use when only qualitative information is required [43].

In addition, the line widths increase with particle size, due to a combination of interband transitions and high-order (electric multi-dipole) plasmon modes, which contribute to increased line widths for larger particles [44]; see Fig. 2.3.

The LSPR properties, such as bandwidth and/or resonance frequency, are sensitive to the size, shape, and dielectric function of the host matrix [45]. Gans theory predicts the LSPR as a function of the aspect ratio and n_d. Under certain conditions, a linear relationship between them can be observed [21]. However, numerical results suggest that, even when the aspect ratio is fixed and the retardation effect

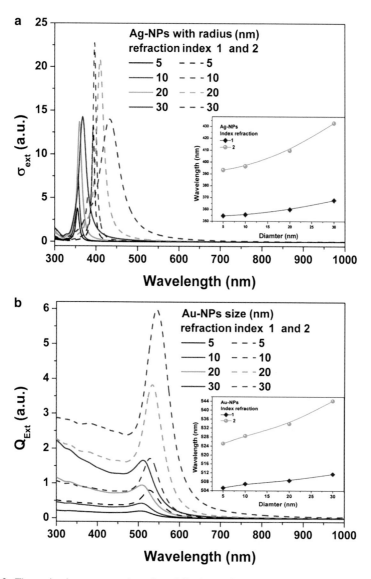

Fig. 2.3 The extinction cross sections from Mie theory for (**a**) Ag and (**b**) Au NPs. Both are functions of the size of the NPs, with a refraction index of 1 (*solid lines*) and 2 (*dotted lines*). The *inset* figures show the red shift with the increasing size of the NPs and the refraction index

is subtle, the λ_{spp} position can still strongly depend on the aspect ratio [46, 47], e.g., in a nanorod or nanowire. By means of the model of Huang et al. [48], we have:

$$\lambda_{\text{spp}} = \pi n_{\text{d}} \sqrt{10\kappa\left(2\delta^2 + r^2\ln\kappa\right)}, \tag{2.16}$$

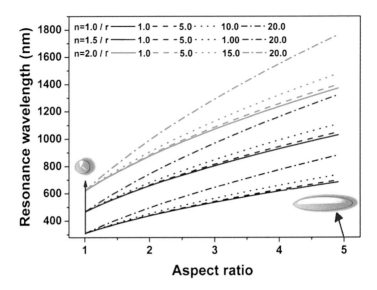

Fig. 2.4 Resonance wavelength as a function of the aspect ratio from Huang et al.'s estimate for radius, with refraction indexes of 1, 1.5, and 2.0. This analysis was realized for an Ag NP

where $\kappa = l/2r$ is the aspect ratio of the NP (where r is the inner radius and l is the cylindrical length), and δ is the skin depth. In this way, λ_{spp} changes with the aspect ratio, κ, of the NP, as in Fig. 2.4. This means a failure of the linear behavior of the oscillating electrons originating from the curved geometry of the NPs, owing to the inertia of the electrons, as in the insets of Fig. 2.3.

From this, particles with sharp rods (nanorods) produce much higher refractive index sensitivities (Fig. 2.4) than would be predicted from nanospheres alone (Fig. 2.3). Over the past decade, a myriad of new NP shapes with ever-increasing refractive index sensitivities have been developed [6, 49, 50].

The λ_{LSPR} peak shift is not strictly linear with changes in n_{d}, as shown in Figs. 2.3 and 2.4. Therefore, the refractive index sensitivity, S, of a particular type of NP is reported in nanometers of peak shift per refractive index unit [49]:

$$S = \frac{d\lambda_{\mathrm{LSPR}}}{dn_{\mathrm{d}}}. \tag{2.17}$$

As LSPR sensing is based on λ_{LSPR} peak shifts, the precision that can be achieved with respect to changes in n_d depends on S and the peak line width. Larger NPs have higher sensitivities, but their spectral shapes are broadened owing to the multipolar excitations and radiative damping. Practically, for a determined material composition, the more red-shifted (lower in energy) that λ_{LSPR} is, then the higher its refractive index sensitivity will be.

2.4   Plasmon–Photon Interaction

Collective charge excitations on the surface of a metallic NP are the origin of LSPR, as we have seen in previous sections. This phenomenon is caused by the incidence of an external EM excitation in the medium in which the NPs are embedded. Since EM radiation is formally explained by quantum mechanics, in which light is represented by "energy packets," or photons, it is reasonable to consider the plasmon as a quantum of energy associated with the collective excitations. This assumption is crucial to comprehending the plasmon–photon conversion process on the NPs, since recent examples in the literature have reported the interaction between plasmonic systems with QEs like QDs [51] and REIs [52, 53], for instance. To describe such interactions, we will use the Green function formalism for the electric potentials, as this has an excellent analytical form for solving the equations that arise from the problem, as has been demonstrated [54].

The spontaneous emission of an isolated QE is due to an external perturbation and is forbidden in the first quantization framework. The situation is completely different if the EM field is also quantized (a second quantization). In the case that QE is placed within a cavity, their interaction is said to be weak. Depending on the characteristics of the cavity, only specific EM modes will be supported, because the cavity can only allow certain final states in which photons can decay. In this manner, if a QE is placed close to a metal NP, then the final density of states will peak at λ_{LSPR}, as it provides a new and strong decay channel on the QE.

This provides us with a strategy for quantifying the interaction between the NP and the QE. In this scenario, we consider two points. (1) If the lifetime of the LSPR is very short compared with the inverse of the spontaneous decay rate of an isolated QE, then we are at the limit of weak coupling. In this case, resonant photons are retained for a short time and interact very little with the emitter. (2) The contrary case occurs when the lifetime of the plasmonic resonance is significantly longer than the spontaneous emission lifetime in free space. Therefore, the plasma–photon interaction is reliable when plasmon is interpreted as a quasi-particle responsible for the collective excitation of the electrons present on the surface of the NP, with a dynamic ruled by a Hamiltonian operator H. Our first task in describing this interaction is to look for a suitable H that governs the system. For simplicity, let us start by considering a particle free from the influence of external fields, assuming only the excitation of free electrons, and neglecting the interband effects from bound electrons (see Chap. 1). Such an assumption fits with the plasma model for a free electron gas that explains the optical properties of a metal. We consider a volume charge that moves into the plasma, defined by its kinetic and potential energy densities [54]:

$$T = \frac{nm}{2}\dot{s}^2 \tag{2.18}$$

and

$$U = \frac{1}{2c^2} j_u^* A^u. \tag{2.19}$$

The term s is the displacement of the volume charge, and consequently $\dot{s}$ is the associated velocity. Furthermore, the term j^u corresponds to the internal sources of the surface of the NP, with charge ρ and current density j, and $A^u = (\phi, A_x, A_y, A_z)$ refers to the scalar and vector potentials of the EM field due to charge fluctuations in the electron gas. The quadrivector potential A^u is evaluated in terms of Green's function:

$$A^u(r,t) = \int G\left(r - r', t' - t\right) j^u\left(r', t'\right) dr' dt'. \tag{2.20}$$

For the EM quadrivector potential, the Green's function satisfies the differential Eq. (2.21):

$$\Box G\left(r, r', t' - t\right) = \delta\left(r - r'\right)\delta\left(t' - t\right), \tag{2.21}$$

where $\Box = \nabla^2 - \frac{1}{c^2}\frac{\partial^2}{\partial t^2}$ is the d'Alembertian operator [63]. The boundary conditions define the Green's function; G must have the follow features: (1) be integrable in $(t' - t)$ and (2) the function vanishes at the surface of the particle. The radiation losses originate from the delay between the quadrivector field $A^u(r, t)$ and the field sources $j^u(r, t)$. This delay explains the dependence of function G on $(t' - t)$. Since we have not considered the influence of interband transitions, there are no ohmic losses in this approximation, i.e., the only loss is due to the aforementioned delay. Using the Green's function notation, it is possible to write the EM potential energy as follows:

$$\int U \, dr = \frac{1}{2c} \int G\left(r, r', t' - t\right) j_u^*(r, t) \cdot j^u\left(r', t'\right) dr dr' dt'. \tag{2.22}$$

Additionally, the field sources can be written in terms of the displacement of the volume charge. The charge density is $j^u = c\rho = -ne\, c\nabla \cdot s$, while the current density is $j^u = j = ne\,\dot{s}$. In both functions, n is the total number of charge carriers per unit volume. Let us assume that the displacement of a volume charge presents a harmonic dependence in time $s(r, t) = s(r)e^{i\omega t}$, similar to an oscillator. In general, the amplitude $s(r)$ is a complex function that does not affect the physical meaning of the final expressions. It is possible to write the potential energy density in terms of the displacement of the charge volume:

$$U = \frac{n^2 e^2}{2c} s^*(r) \cdot \int e^{-i\omega(t'-t)} \left[c^2\nabla_r G\left(r, r', t' - t\right)\nabla_r + c^2\nabla_r^2 G\left(r, r', t' - t\right)\right.$$
$$\left. + \delta\left(r - r'\right)\delta\left(t' - t\right)\right] s(r') dr' dt'. \tag{2.23}$$

To minimize the effects of radiation losses, let us assume that the delay between the field sources and the EM quadrivector occurs only for a short time interval. Therefore, we can rewrite the Green's function of the expression above in terms of a Dirac delta function $(G(r, r', t' - t) \approx G(r, r')\delta(t' - t))$. This approximation eliminates the dependence on time and frequency for the potential energy density. When we substitute it into the argument of the integral above, it is possible to apply the properties of the delta function. This approximation allows us to redefine the amplitudes of the oscillating charges $q_w(r)$ of the NP. Such amplitudes are represented by orthonormal vector functions that fulfill the eigenvalue problem:

$$\omega_w^2 q_w(r) = \frac{ne^2}{m_{NPC}} \cdot \int \left[c^2 \nabla_r G\left(r, r'\right) \nabla_{r'} + c^2 \nabla_r^2 G\left(r, r'\right) + \delta\left(r - r'\right) \right] q_w\left(r'\right) dr'. \quad (2.24)$$

Since the amplitudes were reformulated, the potential energy density is written by:

$$U = \frac{1}{2} n m_{NP} \omega_w^2 q_w^2(r). \quad (2.25)$$

Furthermore, it is possible to define from the amplitude $q_w(r)$ the following transformations:

$$\Psi_w = \frac{1}{\sqrt{V}} \int s(r, t) \cdot q_w(r) dr \quad (2.26)$$

and

$$\Pi_w = \frac{m_{NP}}{\sqrt{V}} \int \dot{s}(r, t) \cdot q_w(r) dr, \quad (2.27)$$

where $s(r, t) = q_w(r)e^{i\omega t}$ is the redefined displacement, m_{NP} and V are the mass and the volume of the NP, respectively. By also applying these results to the kinetic energy density, we are finally able to write the Hamiltonian that describes the dynamics of this system:

$$H = \frac{n}{2} \sum_w \left(\frac{1}{m_{NP}} \Pi_w^2 + m \omega_w^2 \Psi_w^2 \right). \quad (2.28)$$

The collective oscillations of these carriers in a free electron gas generate a current distribution characterized by a set of eigenmodes, q_w. The main consequence of this approach relies on the fact that plasmons are quantized, similar to the well-known case of photons from quantum theory. Each mode is associated with a specific charge displacement $s(r, t)$, which possesses a unique resonance frequency, ω_w; such resonances are responsible for the LSPR effects. Like photons, plasmons are also bosons, because they are defined by a real vector function that corresponds to neutral and massless particles.

Plasmons possess similar features to photons. One of these is the creation/ destruction operators:

$$a_w = \left(\frac{n}{2m_{NP}\hbar\omega_w}\right)^{\frac{1}{2}}(m_{NP}\omega_w\Psi_w + i\Pi_w) \tag{2.29}$$

and

$$a_w^\dagger = \left(\frac{n}{2m_{NP}\hbar\omega_w}\right)^{\frac{1}{2}}(m_{NP}\omega_w\Psi_w - i\Pi_w). \tag{2.30}$$

From these expressions, it is possible to rewrite the operators Ψ_w and Π_w as functions of the creation/destruction operators:

$$\Pi_w = i\left(\frac{m_{NP}\hbar\omega_w}{2n}\right)^{\frac{1}{2}}\left(a_w^\dagger - a_w\right) \tag{2.31}$$

and

$$\Psi_w = \left(\frac{m_{NP}\hbar\omega_w}{2n}\right)^{\frac{1}{2}}\left(a_w^\dagger + a_w\right). \tag{2.32}$$

We are now able to discuss how an EM field produced by an external source interacts with the LSPR from an NP. The interaction of a photon, defined by the wave vector $\boldsymbol{k}$ and polarization α, with a plasmon, which possesses an eigenmode $\boldsymbol{w}$, is expressed in terms of the following Hamiltonian [55]:

$$H = ie\hbar\sqrt{\frac{\pi N}{m_{NP}V}} \cdot \sum_{w,k,\alpha}\sqrt{\frac{\omega_w}{\omega_k}}\Theta_{w,k,\alpha}\left(b_{-k} + b_k^\dagger\right)\left(a_w - a_w^\dagger\right), \tag{2.33}$$

where $(a_w, a_w^\dagger)$ and $(b_k, b_k^\dagger)$ are the creation and destruction operators, respectively, for the plasmon–photon interaction. The coefficient $\Theta_{w,k,\alpha}$ represents all the possible projections of vectorial amplitude of the plasmon with the unit vector u_α that defines the polarization of the photon:

$$\Theta_{w,k,\alpha} = \frac{1}{\sqrt{V}}\int u_\alpha \cdot \boldsymbol{q}_w(\boldsymbol{r})e^{-i\boldsymbol{k}\cdot\boldsymbol{r}}d\boldsymbol{r}. \tag{2.34}$$

Different plasmon modes have different degrees of alignment with the polarization of the external photon, i.e., the coefficient $\Theta_{w,k,\alpha}$ defines how the plasmon modes are aligned with respect to the polarization state of the photon. For instance, Mie theory is an excellent analytical method for determining this coefficient in a nanosphere. Additionally, numerical methods, like FEM and the FDTD method are also a good choice for evaluating this coefficient.

The Hamiltonian operator of Eq. (2.33) defines the plasmon–photon interaction. Thus, it is possible to determine the probabilities of conversion from a plasmon to a photon (and vice versa) using the perturbation method. The first process we will consider is the conversion of a plasmon, initially with a mode w decaying into a photon with a polarization u_α and a wave vector $\mathbf{k}$. The Fermi golden rule defines the probability per unit time for this transition, as follows:

$$T_{w,\mathbf{k},a} = \frac{2\pi}{\hbar} |\langle n_w - 1, 1_{\mathbf{k},\alpha} | H | n_w, 0_{\mathbf{k},\alpha} \rangle|^2 \delta(\hbar\omega_w - \hbar\omega_k). \tag{2.35}$$

By applying the expression for the Hamiltonian in Eq. (2.35), we have the following probability rate:

$$T_{w,\mathbf{k},\alpha} = \frac{2\pi^2 N e^2 \hbar}{m_{NP} V} n_w |\Theta_{w,\mathbf{k},\alpha}|^2 \delta(\hbar\omega_w - \hbar\omega_k). \tag{2.36}$$

Note that this probability rate was achieved from the assumption that the plasmon decays spontaneously. A similar treatment for stimulated emission is also valid. Since the probability rate involves time, it is reasonable to also associate a lifetime, τ_w, for the plasmon on the surface of the NP's surface. A good example of a way to find the lifetime of a plasmon is to consider a spherical metallic NP. This nanosphere presents a diameter d much smaller than the wavelength of the incident radiation, which means that the coefficient $\Theta_{w,\mathbf{k},\alpha}$ does not have any dependence on $\mathbf{K}$, since $kd \ll 1$, as in Eq. (2.34). Consequently, the probability rate T also will not depend on the wave vector. From this approximation, it is possible to obtain the lifetime for a plasmon, since $\tau_w \propto T_w^{-1}$:

$$\tau_w = \frac{2\pi c^3}{V} \omega_0^{-4}, \tag{2.37}$$

where $\omega_0 = \left(\frac{4\pi N e^2}{3m}\right)^{-\frac{1}{2}}$ is the lowest frequency resonance of the plasmon for a metallic nanosphere.

The second process is the conversion of a photon, initially with a wave vector $\mathbf{k}$ and a polarization u_α, into a plasmon defined by its quantum number w. The general form for the probability rate is similar to the first process:

$$T_{w,\mathbf{k},\alpha} = \frac{2\pi^2 N e^2 \hbar}{m_{NP} V} n_{\mathbf{k},\alpha} |\Theta_{w,\mathbf{k},\alpha}|^2 \delta(\hbar\omega_w - \hbar\omega_k). \tag{2.38}$$

To obtain this expression, we assumed that there was no plasmon on the particle before the photon absorption. For both processes reported in this section, we used the perturbation method approximation, which considers only the weak coupling between plasmons in the NP and EM radiation. The increased lifetime of the plasmon is critical to the enhancement of the electric field in the vicinity of the

NP. Therefore, a strong coupling between plasmons and photons must be made. An excellent solution to obtain this strong coupling is to incorporate REIs in the medium in which the metallic NPs are embedded, because additional photons originating from the electronic transitions of these ions will improve the plasmon–photon interaction.

2.4.1 Local Field Enhancement

As discussed above, an SPP is a collective motion of conduction band electrons relative to fixed positive ions in a metal, driven by the electric field component of incident light [39–41, 56, 57]. These oscillations allow for an extraordinary confinement that can generate coupling between neighboring NPs, as in Eqs. (1.32), (1.33), and (2.38). In this manner, the oscillating electric field from the incident light interacts with the delocalized electrons of a metal, such that the electric field acts as a sinusoidal driving force and the Coulombic attraction acts as a restoring force, see Eq. (1.1). Therefore, the electron cloud can be perturbed in such a way that it is physically displaced from the metal framework; see Fig. 2.2. Resonance conditions can be attained when light is coupled in phase with the natural frequency of the SPP. At these resonance conditions (LSPR), the metal structures absorb the maximum amount of incident EM radiation, causing the greatest amount of charge displacement, as in Eqs. (2.23) and (2.25). A consequence of this oscillating dipole moment is that it rapidly dephases. This dephasing time depends on the size and shape of the particle as well as on n_d, and this can be verified in the line width of the plasmon resonance; see Fig. 2.3. In small NPs, the line widths are broadened by electron–surface scattering, and for large particles radiation damping causes broadening [58]. Because the dephasing times are very fast (in some femtoseconds), direct time-resolved measurements of these decay routes are difficult, but not impossible.

An effective coupling between two particles produces a tremendously high local electric field, which may not be possible from individual particles. Fields tend to concentrate at the tips of the protrusions in an effort to be nearly perpendicular to a metallic or semiconductor surface. These areas of intense EM radiation are called "hot spots" [59]. They will also produce intense EM radiation due to a "lightning rod" effect from the high crowding of electric field lines. These effects, namely hot spots and areas of concentrated electric field, are the dominant mechanisms at rough surfaces, and are advantageous for many surface-enhanced spectroscopic methods [60].

The Laplace equation for the potential is $\nabla^2\Phi = 0$, and therefore the electric field is $E = -\nabla\Phi$. The harmonic time dependence can then be added to the solution once the field distributions are identified. Therefore, we can write [61]:

$$E_{\text{in}} = -\frac{3\varepsilon_{\text{d}}}{\varepsilon_{\text{m}} + 2\varepsilon_{\text{d}}}E_0 \tag{2.39}$$

and

$$E_{\text{out}} = E_0 + \frac{3n(n \cdot p) - p}{4\pi\varepsilon_0\varepsilon_d}\left(\frac{1}{r}\right)^3, \tag{2.40}$$

where $p = \varepsilon_0\varepsilon_d\alpha E_0$ is the dipole moment, α is the (complex) polarizability of the NP in the electrostatic approximation, and n is the unit vector in the direction of the point "p" of interest. For NPs with elliptical forms, the polarizabilities α_i along the principal axes ($i = 1, 2,$ and 3) are given by:

$$\alpha_i = \frac{4\pi}{3}a_1a_2a_3\left\{\frac{\varepsilon_m(\omega) - \varepsilon_d(\omega)}{\varepsilon_d(\omega) + L_i(\varepsilon_m(\omega) - \varepsilon_d(\omega))}\right\}, \tag{2.41}$$

where L_i is a geometrical factor given by $L_i = \frac{a_1a_2a_3}{2}\int_0^{\infty}\frac{dq'}{(a_i^2 + q')f(q')}$ and $f(q') = \sqrt{\prod_i^3\left(a_i^2 + q'\right)}$, where $\sum L_i = 1$. For a spherical NP, $L_1 = L_2 = L_3 = 1/3$.

A resonance in α implies a resonant improvement in E_{out}, resulting in exciting applications of NPs to photonic devices. If $|\varepsilon_m + 2\varepsilon_d|$ is minimum, then α_i displays a resonant enhancement. This is similar to the Frohlich condition, Eq. (2.7), which is the associated mode for the NP.

An experimental demonstration of this condition can be found in ref. [4]. In this work, it is shown that Au NPs can be excited by preferential incident radiation, achieving a direct coupling between Er^{3+} ions and Au NPs, resulting in: (1) a local field increase (Frohlich condition), (2) nonradiative decay (heat generation by the Joule effect), or (3) a radiative energy release that depends on the albedo of the NPs. Thus, the response of the NPs will depend on the physical system (e.g., n_d, arrangement) and usually will not be strictly symmetric about the resonance frequency.

Now consider a plane wave incident with $E(r, t) = E_0e^{-i\omega t}$, inducing a $p(t) = \varepsilon_0\varepsilon_d\alpha E_0e^{-i\omega t}$, i.e., we have a scattering of the plane wave by the NP. The EM fields associated with an ED in the near, intermediate, and radiation zones are [61]:

$$H = \frac{ck^2}{4\pi}(n \times p)\frac{e^{ik\cdot r}}{r}\left(1 - \frac{1}{ik \cdot r}\right) \tag{2.42}$$

and

$$E = \frac{e^{ik\cdot r}}{4\pi r\varepsilon_0\varepsilon_d}\left\{k^2(n \times p) \times n + (3n(n \cdot p) - p)\left(\frac{1}{r^2} - \frac{ik}{r}\right)\right\}. \tag{2.43}$$

For $kr \ll 1$ (near zone), we have Eq. (2.43), and the magnetic field is $\propto n \times p/r^2$. Therefore, in the near field, the fields are mainly electric, and for the static field $k \cdot r \to 0$, the magnetic field vanishes. If $k \cdot r \gg 1$, then the dipole fields have spherical-wave forms:

$$H = \frac{ck^2}{4\pi}(n \times p)\frac{e^{ik \cdot r}}{r} \tag{2.44}$$

and

$$E = \sqrt{\frac{\mu_0}{\varepsilon_0 \varepsilon_d}} H \times n. \tag{2.45}$$

The improvement obtained with these NPs is due to the formation of EDs, which generates a polarization. Therefore, a modification in the local electric field is produced by those ED that can be assumed as a local field correction [62]:

$$E_{eff} = \frac{(\varepsilon_0 + 2)E_0}{3}\left(1 + V\omega_p{}^2\left\{(1-V)\left(\frac{\omega_p{}^2}{3\varepsilon_0}\right) - \omega^2 + i\gamma\omega\right\}^{-1}\right), \tag{2.46}$$

where E_{eff} is the effective electric field. In the presence of an EM field, we have LSPR, which forms EDs separated by diverse distances between them or the QEs, some of which will contribute to the improvement in luminescence. The effective field is found to be: $E_{eff} = E_{loc} + E$. Consequently, the enhancement luminescence intensity is related to the strong local electric field owing to the NPs, which increases the quantum yield η of the luminescence, defined as [63]:

$$\eta = \frac{r+d}{d} = \frac{|E_{loc}|}{|E|}, \tag{2.47}$$

where $d (=2R)$ is the NP size and r is distance between two NPs or NP and QE. The maximum field is given by the shortest distance between two equipotential particles. Both the "hot spot" and the "lightning rod" effects are also consequences of the short distance, see Eq. (2.47), between two equipotential particles (e.g., a REI coupled with an NP), see Fig. 2.5. Hence, for small separations, it becomes essential to adapt a quantum mechanical description in order to predict reliable electric field enhancements, as discussed in Sect. 2.4.

The NP will radiate at the same frequency ω as the applied field with intensity:

$$I_{NP}(\omega) = \frac{P^2}{32\pi^2 c^3 \epsilon_0}\int_0^{2\pi} \sin^2(\theta)d\theta = \frac{|P|^2\omega^4}{12\pi c^3 \epsilon_0}. \tag{2.48}$$

In the classical Lorentzian model, the dielectric function for a single oscillator of mass m is given by Eq. (1.24). However, this equation makes no assumptions about the long-range order of the oscillator. Glass materials appear to possess no long-range order, as their coordination numbers are usually the same and their bond lengths, bond angles, and resonance frequencies are very close to those of their crystalline counterparts. However, this can be an approximation, as amorphous materials are very complex and almost certainly vary from sample to sample. In this scenario, we will assume that glass materials can be described as a mixture

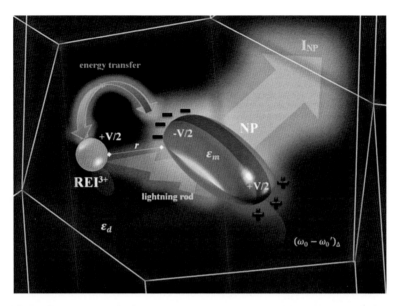

Fig. 2.5 Possible processes for interaction between an NP and REI within an amorphous net, determined by the distance between two equipotential particles, where $I_{NP}(\omega)$ is defined by Eq. (3.10)

(distribution) of N different oscillators, i.e., we can assume a dielectric function as a continuous linear superposition of individual dielectric functions [64, 65]:

$$\varepsilon(\omega) = \varepsilon_0 + \int \prod_\Delta \left(\omega_0 - \omega_0'\right) \frac{\omega_i^2}{\omega_0'^2 - \omega^2 - i\gamma_0\omega} d\omega_0', \qquad (2.49)$$

where $\varepsilon(\omega)(=\varepsilon_d(\omega))$ is the dielectric function of the host matrix (glass) and $\prod(\omega_0 - \omega_0')$ is a rectangular function of width Δ that represents the relative numbers of oscillators at each frequency from the transparent material. Now, we have:

$$I_{NP}(\omega) = \frac{4\pi\omega^4 |E|^2 (a_1 a_2 a_3)^2}{9c^3\varepsilon_0}$$

$$\times \left\{ \frac{1 - \dfrac{\omega_p^2}{\omega^2 + i\gamma\omega} - \left(\varepsilon_0 + \displaystyle\int \prod \left(\omega_0 - \omega_0'\right) \dfrac{\omega_i^2}{\Delta\omega_0'^2 - \omega^2 - i\gamma_0\omega} d\omega_0'\right)}{\left(\varepsilon_0 + \displaystyle\int \prod \left(\omega_0 - \omega_0'\right) \dfrac{\omega_i^2}{\Delta\omega_0'^2 - \omega^2 - i\gamma_0\omega} d\omega_0'\right)(1 - L_i) + L_i \left(1 - \dfrac{\omega_p^2}{\omega^2 + i\gamma\omega}\right)} \right\}^2.$$

$$(2.50)$$

This is a complete equation that describes the emission intensity of the NP as a function of $\omega(I_{NP}(\omega))$, which considers the geometrical effect of the NP and the environment around as a transparent material. Furthermore, we consider the following cases. (1) If $\varepsilon_m(\omega) - \varepsilon_d(\omega) > 0$, then $I_{NP}(\omega) > 0$ is always the case. Therefore, we have an improvement in the luminescence, as mentioned above. (ii) If $\varepsilon_m(\omega) - \varepsilon_d(\omega) < 0$, then we most likely (based on a geometrical dependency, $L_i \gg 1$) have a quenching in the luminescence. This reflects the asymmetric behavior of the NP emissions ($I_{NP}(\omega)$) for different frequencies; this can be verified in the experimental results [6]. Figure 2.5 shows the effects discussed here.

2.4.2   Energy Transfer

Generally speaking, an energy transfer process occurs between two (or more) QEs, where one of these (the donor, which is in an excited state with more energy than the other QEs) transfers part of its energy to other QE(s) (acceptors, in an excited state of less energy) via an interaction between two induced dipoles. These can either be between ions of the same rare earth (clusters), or between different ions (the sensitization of one REI by another). The energy transfer efficiency depends on the inverse of the distance r between the donor and the acceptor. This process is widely used in studies of structural and dynamic information in macromolecular assemblies (biology, biochemistry, and polymer science), allowing one to determine the long-range distance [66, 67]. However, due to the context of our book, we will focus on talking about the effects of this process in REIs and NPs.

The energy transfer process may occur between REIs of the same or different species, and it may be either favorable or detrimental, which can be verified through its fluorescence/luminescence emission spectrum. Engineering of this energy transfer process allows, for example, improving the pumping efficiency of a solid state laser [68], increasing the bandwidth of band emission of ions in the optical telecommunications band [69–72], tuning of the emission color of these ions in the visible region [73–75] up to and including white light [76–78], and improving the performance of a solar cell [79]. However, the energy transfer process is a dissipative mechanism that decreases the emission intensity or lifetime of these ions [80]. The study of luminescence from REIs has been a subject of steady growth over the past few decades, principally owing to the ever-increasing demand for optical sources and amplifiers operating at wavelengths compatible with fiber communications technology, diagnostic images, optical displays, and solar energy. Specifically, the interest in luminescent REIs has concentrated on Er^{3+} and, in particular, its emission band around 1.54 μm. The reason for this is plain to see if one considers the rapid growth in optical communications and some of the limitations of materials used in this technology [81].

We must say that the $4f$ electrons of the trivalent REI exhibit a multiplicity of excited states in both emission and absorption, as in Sect. 1.4. Then, the emission/absorption spectra resulting from transitions between the various levels of these

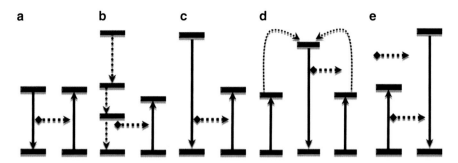

Fig. 2.6 Energy transfer processes for (**a**) multipolar resonance, (**b**) multipolar transfer, (**c**) nonresonant transfer interactions, (**d**) stepwise up-conversion, and (**e**) cooperative luminescence

states consist of groups of sharp lines. Therefore, energy transfer here can be described as interactions that result in the transfer of excitation energy between ions (de-excitation transitions), resulting in radiative (emission and radiative transfer) and nonradiative (internal relaxation and multipolar interaction) processes between ions [82].

Multipolar interactions may be divided into five categories: (a) multipolar resonance, (b) multipolar transfer, (c) nonresonant transfer interactions, (d) stepwise up-conversion, and (e) cooperative luminescence, as shown in Fig. 2.6.

These interactions are described using either the short-range exchange or longer-range electric multipolar mechanisms. Multipolar interactions between neighboring REIs are a topic of continuing study; the first detailed treatments were given by Forster [83], Dexter [84], and Van Uitert (specifically on REIs) [82]. In these models, two nearby ions are called the donor (D) and acceptor (A). The D is in the excited state, and the A is initially unexcited. Note that the two ions need not be both of the same RE. Therefore, the transfer efficiency may be defined as:

$$\eta_D = \frac{W_{DA}\tau_D}{1 + W_{DA}\tau_D}, \qquad (2.51)$$

where W_{DA} is the rate of the transfer process and τ_D is the radiative lifetime of the D ion in the absence of an A ion. In fact, such ion–ion interaction reduces the observed luminescence lifetime:

$$\tau_{obs} = \tau_{D+A} = \tau_D(1 - \eta_D). \qquad (2.52)$$

Dexter's theory assumes an interacting ion pair, so by integrating over all such pairs in a macroscopic sample, we have:

$$\widetilde{\eta}_D = \int_0^\infty \eta_D P(A) dV', \qquad (2.53)$$

where $P(A)$ is the probability of finding an A ion at a distance from the D between r and $r + \Delta r$, and V' is the interaction volume $(= 4\pi r^3/3)$. If ρ_A is the acceptor density, then we have $P(A)dV' = \rho_A \left(e^{\rho_A V'} \right)$. Then:

$$W_{DA}\tau_D = \left(\frac{1}{\rho_A V'} \right)^{n/3}. \tag{2.54}$$

In this manner, we have the $1/r^x$ dependence of the strength of the multipolar interactions, which yields: $x = 6, 8$, and 10, i.e., dipole–dipole, dipole–quadrupole, and quadrupole–quadrupole interactions, respectively. Nonradiative energy transfer can effectively cover much longer distances. Nevertheless, REI transitions strongly depend on the environment (i.e., the crystal field, see Chap. 1, Sect. 1.5.1.2). Notice that W_{DA} can be altered owing to the presence of NPs, because this produces a change in the local electric field local; see Fig. 2.5.

Lakowicz [85] defined this W_{DA} as:

$$W_{DA}(r) = \frac{0.53o^2}{n_d{}^2 N \tau_D r^6} J(\lambda), \tag{2.55}$$

where $\square$ is the quantum yield of the D in the absence of A and N is Avogadro's number. The term o^2 is a factor describing the relative orientation in space of the transition dipoles of the D and A, and is usually assumed to be 2/3 for dynamic random averaging of the D and A. In fact, the only case in which energy transfer is necessarily forbidden is when the ED from the D and A are orthogonal to r, and in this situation $o^2 = 0$. Meanwhile, $J(\lambda)$ is defined as:

$$J(\lambda) = \frac{\displaystyle\int_0^\infty F_D(\lambda)\epsilon_A(\lambda)\lambda^4 d\lambda}{\displaystyle\int_0^\infty F_D(\lambda)d\lambda}, \tag{2.56}$$

where $F_D(\lambda)$ is the corrected luminescence intensity of the D in the λ to $\lambda + \Delta\lambda$ region, with the total intensity (area under the curve) normalized to unity. Next, $\epsilon_A(\lambda)$ is the extinction coefficient of the A in the λ to $\lambda + \Delta\lambda$ region. Therefore, the overlap integral $J(\lambda)$ expresses the degree of spectral overlap between the D emission and the A absorption; see Fig. 2.7.

As mentioned above, the energy transfer process can be meaningfully modified by interaction with the NP. For example, NPs within the dielectric medium can modify the bandwidth of the Stark energy levels, which results in a blue or red shift and or a broadening or increase in the REI emissions [6]. At a simple level, the influence can be understood as a consequence of effecting shifts in the electron distributions of the interacting REI, producing modified transition moments [5].

We can use the dynamic coupling mechanism [86–88] and show that the interaction energy between the NP and the REI is given by:

Fig. 2.7 Energies and
spectral overlap features for
forward and backward
energy transfer between the
D and A

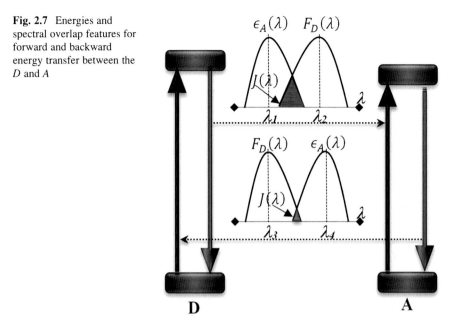

$$H_{DC} = e \sum_{i,j} \boldsymbol{p}_{NPj} n_j \frac{\boldsymbol{r}_{ij}}{|\boldsymbol{r}_{ij}|^3}, \qquad (2.57)$$

where $\boldsymbol{p}_{NPj}(=\alpha_j \boldsymbol{E})$ is the ED moment of the NP induced by the incident light, $\boldsymbol{r}_{ij}$ is the separation distance between the NP and REI, n_j is the density of the conduction electrons inside a plasma characterized by carriers with charge $N'e$ analogous to s $(\boldsymbol{r},t)$ in Sect. 2.4, and α_j was defined in Eq. (2.41). The probability rate for the NP is given by Eq. (2.38). The Hamiltonian H_{DC} is expressed as a function of the set of eigenmodes associated with the conduction electron density generated by different NPs (j) through its collective free oscillations at each resonance frequency of the NP and their geometric dependence by means of α_j. The quantum behavior of the REI and NPs (in the LSPR mode) interacts through H_{DC}, resulting in a plasmon–photon coupling, which can be verified as an enhancement/quenching of luminescence, a widening of the broadband emission, and a change in its line shape [4–6, 32, 62, 63, 89]. Such interaction, as described in Sect. 2.4, is given by n_j, where a charge transfer mechanism (resonant or nonresonant, as in Fig. 2.6) between the NP and REI only occurs if the REI is close enough to the NP for the wave functions of the two systems to overlap, as shown in Figs. 2.5 and 2.7.

From the formulation of coupling efficiency, $\eta_{LSPR} = \Gamma_{LSPR}/(\Gamma_{rad} + \Gamma_{nonrad} + \Gamma_{LSPR})$, where Γ_{LSPR} is the energy transfer rate to the LSPR mode [90, 91]. Therefore, the total intensity I_T can be written as [6]:

$$I_{D+A} = I_D(1 - \eta_{LSPR}) + \eta_0 I_2. \tag{2.58}$$

Here, I_D is the intensity emission of the REI and $I_2 (=I_A)$ is the intensity emission of the NP; see Eq. (2.50), at equipotential surfaces with electric potentials $-V/2$ and $V/2$ for the NP and the REI, respectively, i.e., ED coupling. In the case with multiple interactions at different distances between NPs:REIs, the charge from the REI is the same. Because the $\vec{E}_{inc}$ employed to excite these particles will be absorbed first for the REI due to the cross section absorption those REI is more than of the NPs, Fig. 2.8. Case contrary the light incident ($\lambda_0 \neq \lambda_{spp}$—wavelength surface plasmon resonance) is scattered for the NPs, and this scattered light is absorbed for the REI which emits and this coupled/resonates with the NP. Finally, η_0 is the internal quantum efficiency of the energy transfer. Note that here we did not consider the shape, size, or material of the NP, Fig. 2.8.

Experimentally, we can calculate the energy transfer probability as [92]: $P_{D-A} = \frac{1}{\tau_D}\left(\frac{I_D}{I_{D+A}} - 1\right)$. Let assume us that: $P_{D-A} = \eta_{LSPR}$, we can write:

$$I_T - I_D = \eta_0 I_2 - I_D \eta_{LSPR} \tag{2.59}$$

and

$$\eta_0 I_2 = I_T - I_1\left(1 - \frac{1}{\tau} + \frac{I_D}{I_{D+A}\tau}\right). \tag{2.60}$$

Therefore, if $I_2 > 0$, then we have an improvement in the luminescence intensity, and otherwise we have quenching. This is a similar case to that obtained in the equation of the energy transfer probability, i.e., if $\frac{I_D}{I_{D+A}} - 1 < 1$, then we have an improvement in the luminescence, and otherwise a quenching.

In order to demonstrate the application of this model (see Eqs. (2.59) and (2.60)), we will use the emission spectra of our sample of Er^{3+}-doped tellurite glass with Au NPs embedded, under thermal treatments for 2 and 12 h. Glasses of $74TeO_2$-$5ZnO$-$15Na_2O$-$5GeO_2$-$1Er_2O_3$:$0.05AuCl_3$ were prepared by the traditional melt-casting technique. More details about the preparation are provided in a previous paper [72]. The lifetime of the sample without gold NPs is 18.5 µs. It is centered at 545 nm under the excitation of a laser at 488 nm.

Figure 2.9 shows the curves obtained after evaluation of the model and for both quenching and enhancement. In both cases, the behavior can be understood as $I_{NP} = I_{NP}(\omega)$, see Eq. (2.50). The energy transfer from the REI → NP is more probable than the energy transfer from the NP → REI, since the excited state lifetime of NPs is extremely short (on the order of nanoseconds) in comparison with the higher energy excited states of the Er^{3+} ions (on the order of microseconds), as shown in Fig. 2.9a. However, the luminescence enhancement is attributed to a local field enhancement, thus LSPR gives us the ability to significantly alter the "local density of optical states" (LDOS). It is well known that the

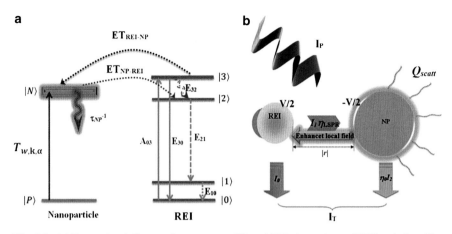

Fig. 2.8 (a) Energy level diagram for a resonant NP and REI absorption and REI emission. The energy transfers between RE → NP or NP → RE are represented by *dotted lines*, and the *vertical dotted line* displays the radiative transition under consideration. The *curved arrows* indicate nonradiative transitions. (b) A schematic representation of the REI–NP interaction. A plane wave with pump intensity I_1 causes the following processes: (1) absorption of the REI, I_p; (2) LSPR from NPs, $I_D\eta_{LSPR}$; and (3) an NP transmitter. Here the coupling depends on η_{LSPR}

spontaneous emission of REIs is proportional to the LDOS, resulting in an enhanced emission resulting from the changes in the LDOS of the REI or other QEs, as in Fig. 2.9b.

In addition, the REI–NP interaction waves will result in constructive and destructive interference phenomena, since both particles are located very close to each other (an effect observed in the far field in Fig. 2.9), corresponding to a maximum ($I_T > 0$) and a minimum ($I_T < 0$) of the emission spectra. Therefore, these asymmetric line shapes (obtained from our model) can be considered as quantum interference, and are produced from the rapid variations in intensity of the various diffracted spectral orders of REIs and NPs embedded within the tellurite glass. Similar asymmetric profiles were observed in various other systems and settings [93]. In this sense, these experimental results demonstrate for the first time one of the basic properties of Fano resonance between REI and metallic NP, i.e., resonances with constructive (luminescence enhancement) or destructive (luminescence quenching) interference.

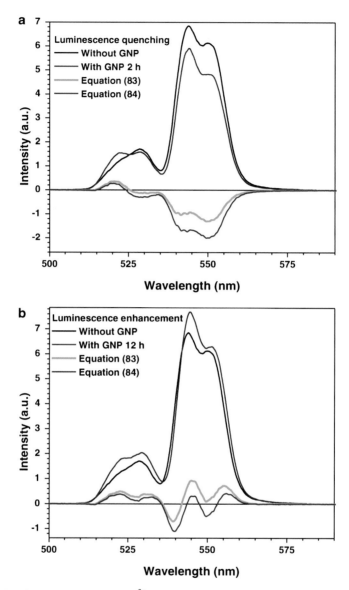

Fig. 2.9 Luminescence spectra from Er^{3+}-doped tellurite glass with gold NPs (GNP) excited with a diode laser at 488 nm. (**a**) We observe a quenching of the emission intensity due to the presence of GNPs. The glass has a thermal treatment of 2 h. (**b**) Luminescence enhancement due to the presence of GNPs. The glass has a thermal treatment of 12 h. Both *curves* were fitted with Eqs. (2.59) and (2.60). Here, the erbium ion is the donor ion and the GNP is the acceptor particle (**a**) or vice versa (**b**)

2.5 Conclusion

The research into interacting plasmonic NPs has grown in scientific and technological importance and particularly allows us manipulate the quantum states of a quantum emitter and control its EM radiation at the nanoscale, producing innumerable practical applications in spectroscopic techniques and improved images.

In this chapter, we reviewed the fundamental concepts, recent advances, and applications related to the interaction of NPs with light and quantum emitters, especially with REIs. The different dynamical processes in metallic NPs, dephasing of LSPR, electron–electron scattering, electron–photon coupling, lightning rods, coherent excitation, energy transfer, and damping of vibrational modes, all depend on the size, shape, and composition of the particles and the environment of the medium, and they are well detailed both theoretically and experimentally.

There are many theoretical and experimental techniques available to study plasmonic NPs and their properties, and researchers are currently looking to extend the capacity of these. Thus, we can exploit many of the abilities of new nanodevices based on plasmonics with important applications.

From these NPs, it is possible to modulate the down/up-conversion emission of the REIs with applicability in areas such as optical communication, including energy conversion and biomedical imaging or the exploitation of coherent harmonic generation processes from plasmonic NPs excited at resonance. This enhanced fluorescence has high potential for applications in photonics, such as lasers, optical displays, and optical memory devices. The success of new applications of NPs is subject to improvement in the understanding of the properties of LSPR and the surrounding environment. Hence, exhaustive studies must be realized in order to provide further progress in REI–NP interaction over a broad spectrum of applications. Additionally, our experimental results show for the first time (Fig. 2.9) one of the basic properties of Fano resonance—quantum interference.

References

1. Novotny, L., Hecht, B. (eds.): Principles of Nano-Optics. Cambridge University Press, Cambridge (2006). Chapter 9
2. Utikal, T., Eichhammer, E., Petersen, L., Renn, A., Gotzinger, S., Sandoghdar, V.: Spectroscopic detection and state preparation of a single praseodymium ion in a crystal. Nat. Commun. 5, 3627 (2014)
3. Thiel, C.W., Bottger, T., Cone, R.L.: Rare-earth-doped materials for applications in quantum information storage and signal processing. J. Lumin. 131(3), 353 (2011)
4. Rivera, V.A.G., Ledemi, Y., Osorio, S.P.A., Manzani, D., Ferri, F.A., Ribeiro, S.J.L., Nunes, L.A.O., Marega Jr., E.: Tunable plasmon resonance modes on gold nanoparticles in Er^{3+}-doped germanium–tellurite glass. J. Non-Cryst. Solids 378, 126 (2013)
5. Rivera, V.A.G., Osorio, S.P.A., Ledemi, Y., Manzani, D., Messaddeq, Y., Nunes, L.A.O., Marega Jr., E.: Localized surface plasmon resonance interaction with Er^{3+}-doped telluriteglass. Opt. Exp. 18(24), 25321 (2010)

6. Rivera V.A.G., Ferri, F.A., Marega, E. Jr.: In: Kim KY (ed.) Localized Surface Plasmon Resonances: Noble Metal Nanoparticle Interaction with Rare-Earth Ions, Chapter 11, Intech, Croatia (2012).
7. Tame, M.S., et al.: Quantum plasmonics. Nat. Phys. **9**, 329 (2013)
8. Nature Photonics: Special Issue on Plasmonics. **6**(11) 707–794 (2012)
9. Bharadwaj, P., Deutsch, B., Novotny, L.: Optical antennas. Adv. Opt. Photon. **1**, 438 (2009)
10. Lim, Z.Z.J., Li, J.E.J., NG, C.T., Yung, L.Y.L., Bay, B.H.: Gold nanoparticles in cancer therapy. Acta. Pharmac. Sinica **32**, 983 (2011)
11. Zhang, Z., Wang, L., Wang, J., Jiang, X., Li, X., Hu, Z., Ji, Y., Wu, X., Chen, C.: Mesoporous silica-coated gold nanorods as a light-mediated multifunctional theranostic platform for cancer treatment. Adv. Mater. **24**(11), 1418 (2012)
12. Mayer, K.M., Hao, F., Lee, S., Nordlander, P., Hafner, J.H.: A single molecule immunoassay by localized surface plasmon resonance. Nanotechnology **21**, 255503 (2010)
13. Sepulveda, B., Angelmore, P.C., Lechuga, L.M., Marzan, L.M.L.: LSPR-based nanobiosensors. Nano Today **4**(3), 244 (2009)
14. Giannini, V., Dominguez, A.I.F., Heck, S.C., Maier, S.A.: Plasmonic nanoantennas: Fundamentals and their use in controlling the radiative properties of nanoemitters. Chem. Rev. **111**(6), 3888 (2011)
15. Michel, J.F.: Digonnet, Rare-Earth-Doped Fiber Laser and Amplifiers, 2nd edn. Marcel Dekker InC, New York, NY (2001)
16. Yang, W.-H., Schatz, G.C., Duyne, R.P.V.: Discrete dipole approximation for calculating extinction and Raman intensities for small particles with arbitrary shapes. J. Chem. Phys. **103**, 869 (1995)
17. Yee, K.: Numerical solution of initial boundary value problems involving Maxwell's equations in isotropic media. IEEE Trans. Antennas. Propagat. **14**, 302 (1966)
18. Jin, J.: The Finite Element Method in Electromagnetics. Wiley, New York, NY (2002)
19. Mayer, K.M., Hafner, J.H.: Localized surface plasmon resonance sensors. Chem. Rev. **111**, 3828 (2011)
20. Kelly, K.L., Coronado, E., Zhao, L.L., Schatz, G.C.: The optical properties of metal nanoparticles: the influence of size, shape, and dielectric environment. J. Phys. Chem. B **107**, 668 (2003)
21. Link, S., Mohamed, M.B., El-Sayed, M.A.: Simulation of the optical absorption spectra of gold nanorods as a function of their aspect ratio and the effect of the medium dielectric constant. J. Phys. Chem. B **103**, 3073 (1999)
22. Miller, M.M., Lazarides, A.A.: Sensitivity of metal nanoparticle surface plasmon resonance to the dielectric environment. J. Phys. Chem. B **109**, 21556 (2005)
23. Dai, D., He, S.: A silicon-based hybrid plasmonic waveguide with a metal cap for a nano-scale light confinement. Opt. Exp. **17**(19), 16646 (2009)
24. Ferri, F.A., Rivera, V.A.G., Osorio, S.P.A., Silva, O.B., Zanatta, A.R., Borges, B.H.V., Weiner, J., Marega Jr., E.: Influence of film thickness on the optical transmission through subwavelength single slits in metallic thin films. Appl. Opt. **50**(31), G11 (2011)
25. Ferri, F.A., Rivera, V.A.G., Silva, O.B., Osorio, S.P.A., Zanatta, A.R., Borges, B.-H.V., Weiner, J., Marega Jr., E.: Surface plasmon propagation in novel multilayered metallic thin films. Proc. SPIE **8269**, 826923 (2012)
26. Sun, Z., Jung, Y.S., Kim, H.K.: Role of surface plasmons in the optical interaction in metallic gratings with narrow slits. Appl. Phys. Lett. **83**(15), 3021 (2003)
27. Catchpole, K.R., Polman, A.: Design principles for particle plasmon enhanced solar cells. Appl. Phys. Lett. **93**, 191113 (2008)
28. Acimovic, S.S., et al.: Plasmon near-field coupling in metal dimers as a step toward single-molecule sensing. ACS Nano **3**(5), 1231 (2009)
29. Mallidi, S., et al.: Multiwavelength photoacoustic imaging and plasmon resonance coupling of gold nanoparticles for selective detection of cancer. Nano Lett. **9**(8), 2825 (2009)

30. Lu, X., Rycenga, M., Skrabalak, S.E., Wiley, B., Xia, Y.: Chemical synthesis of novel plasmonic nanoparticles. Annu. Rev. Phys. Chem. **60**, 167 (2009)
31. Blaber, M.G., Arnold, M.D., Ford, M.J.: Search for the ideal plasmonic nanoshell: the effects of surface scattering and alternatives to gold and silver. J. Phys. Chem. C **113**(8), 3041 (2009)
32. Osorio, S.P.A., Rivera, V.A.G., Nunes, L.A.O., Marega Jr., E., Manzani, D., Messaddeq, Y.: Plasmonic coupling in Er^{3+}:Au tellurite glass. Plasmonics **7**(1), 53 (2012)
33. Zori, I., Zach, M., Kasemo, B., Langhammer, C.: Gold, platinum, and aluminum nanodisk plasmons: Material independence, subradiance, and damping mechanisms. ACS Nano **5**(4), 2535 (2011)
34. Langhammer, C., Schwind, M., Kasemo, B., Zoric, I.: Localized surface plasmon resonances in aluminum nanodisks. Nano Lett. **8**, 1461 (2008)
35. Knight, M.W., King, N.S., Liu, L., Everitt, H.O., Nordlander, P., Halas, N.J.: Aluminum for plasmonics. ACS Nano **8**(1), 834 (2014)
36. George, H.C., Zhao, J., Schatz, G.C., Van Duyne, R.P.: Localized surface plasmon resonance spectroscopy of triangular aluminum nanoparticles. J. Phys. Chem. C **112**(36), 13958 (2008)
37. DeSantis, C.J., Weiner, R.G., Radmilovic, A., Bower, M.M., Skrabalak, S.E.: Seeding bimetallic nanostructures as a new class of plasmonic colloids. J. Phys. Chem. Lett. **4**, 3072 (2013)
38. Chemical Reviews: Special issue on plasmonics. **111**(6), 3667–3994 (2011)
39. Quinten, M.: Optical Properties of Nanoparticles Systems. Wiley-VCH Verlag GmbH & Co. KGaA, Weinheim (2011)
40. Fendler, J.H. (ed.): Nanoparticles and Nanostructured Films. Wiley-VCH Verlag GmbH & Co. KGaA, Weinheim (1998)
41. Maier, S.A.: Plasmonics: Fundamentals and Applications. Springer Science + Business Media LLC, New York, NY (2007)
42. Palik, E.D.: Handbook of Optical Constants of Solids II. Elsevier, Orlando, FL (1998)
43. Lance Kelly, K., Coronado, E., Zhao, L.L., Schatz, G.C.: The optical properties of metal nanoparticles: the influence of size, shape, and dielectric environment. J. Phys. Chem. B **107**, 668 (2003)
44. Link, S., El-Sayed, M.A.: Size and temperature dependence of the plasmon absorption of colloidal gold nanoparticles. J. Phys. Chem. B **103**, 4212 (1999)
45. Evanoff, D.D., White, R.L., Chumanov, G.: Measuring the distance dependence of the local electromagnetic field from silver nanoparticles. J. Phys. Chem. B **108**(37), 1522 (2004)
46. Kuwata, H., Tamaru, H., Esumi, K., Miyano, K.: Resonant light scattering from metal nanoparticles: practical analysis beyond Rayleigh approximation. Appl. Phys. Lett. **83**, 4625 (2003)
47. Prescott, S.W., Mulvaney, P.: Gold nanorod extinction spectra. J. Appl. Phys. **99**, 123504–123510 (2006)
48. Huang, C.P., Yin, X.G., Huang, H., Zhu, H.Y.: Study of plasmon resonance in a gold nanorod with an LC circuit model. Opt. Exp. **17**(8), 6407 (2009)
49. Mayer, K.M., Hafner, J.H.: Localized surface plasmon resonance sensors. Chem. Rev. **111**, 3828 (2011)
50. Rycenga, M., Cobley, C.M., Zeng, J., Li, W., Moran, C.H., Zhang, Q., Qin, D., Xia, Y.: Controlling the synthesis and assembly of silver nanostructures for plasmonic applications. Chem. Rev. **111**, 3669 (2011)
51. Ridolfo, A., et al.: Quantum plasmonics with quantum dot-metal nanoparticle molecules: influence of the fano effect on photon statistics. Phys. Rev. Lett. **105**, 263601 (2010)
52. Silva, O.B., Rivera, V.A.G., Ferri, F.A., Ledemi, Y., Zanatta, A.R., Messaddeq, Y., Marega Jr., E.: Quantum plasmonic interaction: Emission enhancement of Er^{3+}-Tm^{3+} co-doped tellurite glass via tuning nanobowtie. Proc. SPIE **8809**, 88092X–1 (2013)
53. Rivera, V.A.G., Ledemi, Y., Osorio, S.P.A., Manzani, D., Messaddeq, Y., Nunes, L.A.O., Marega Jr., E.: Efficient plasmonic coupling between Er^{3+}:(Ag/Au) in tellurite glasses. J. Non-Cryst. Solids. **358**(2), 399 (2012)

54. Finazzi, M., Ciaccacci, F.: Plasmon-photon interaction in metal nanoparticles: second-quantization perturbative approach. Phys. Rev. B **86**, 035428 (2012)
55. Sakurai, J.J.: Advanced Quantum Mechanics. Addison-Wesley, New York, NY (1967)
56. Schmid, G. (ed.): Nanoparticles. Wiley-VSH Verlag GmbH & Co. KGaA, Boschstr (2010)
57. Raether, H.: Surface Plasmon on Smooth and Rough Surfaces and on Gratings, Springer Tracts in Modern Physics, vol. 111. Springer, New York, NY (1988)
58. Hu, M., Novo, C., Funston, A., Wang, H.N., Staleva, H., Zou, S.L., Mulvaney, P., Xia, Y.N., Hartland, G.V.: Dark-field microscopy studies of single metal nanoparticles: understanding the factors that influence the linewidth of the localized surface plasmon resonance. J. Mater. Chem. **18**, 1949 (2008)
59. Shalaev, V.M., Botet, R., Jullien, R.: Resonant light scattering by fractal clusters. Phys. Rev. B **44**, 12216 (1991)
60. Haes, A.J., Haynes, C.L., McFarland, A.D., Schatz, G.C., Van Duyne, R.P., Zou, S.L.: Plasmonic materials for surface-enhanced sensing and spectroscopy. MRS Bull. **30**, 368 (2005)
61. Jackson John, D.: Classical Electrodynamics, 3rd edn. Wiley, New York, NY (1999)
62. Malta, O.L., Santa-Cruz, P.O., de Sa, G.F., Auzel, F.: Fluorescence enhancement induced by the presence of small silver particles in Eu^{3+}-doped materials. J. Lumin. **33**, 261 (1985)
63. Som, T., Karmakar, B.: Nanosilver enhanced upconversion fluorescence of erbium ions in Er^{3+}:Ag-antimony glass nanocomposites. J. Appl. Phys. **105**, 013102 (2009)
64. Lynch, D.K.: A new model for the infrared dielectric function of amorphous materials. Astrophys. J. **467**, 894 (1996)
65. Bohren, C.F., Huffman, D.R.: Absorption and Scattering of Light by Small Particles. Wiley, New York, NY (1983)
66. Pengguang, W., Brand, L.: Resonance energy transfer: methods and applications. Anal. Biochem. **218**, 1 (1994)
67. Clegg, R.M.: Fluorescence resonance energy transfer. Curr. Opin. Biotech. **6**, 103 (1995)
68. Snitzer, E., Woodcock, R.: Yb^{3+}-Er^{3+} glass laser. Appl. Phys. Lett. **6**, 45 (1965)
69. Milanese, D., Vota, M., Chen, Q., Xing, J., Liao, G., Gebavi, H., Ferraris, M., Coluccelli, N., Taccheo, S.: Investigation of infrared emission and lifetime in Tm-doped $75TeO_2$:20ZnO:5Na_2O (mol%) glasses: Effect of Ho and Yb co-doping. J. Non-Crys. Solids **354**, 1955 (2008)
70. Pal, A., Dhar, A., Das, S., Chen, S.Y., Sun, T., Sen, R., Grattan, K.T.V.: Ytterbium-sensitized Thulium-doped fiber laser in the near-IR with 980 nm pumping. Opt. Exp. **18**(5), 5068 (2010)
71. Zhou, B., Tao, L., Tsang, Y.H., Jin, W., Pun, E.Y.B.: Superbroadband near-IR photoluminescence from Pr^{3+}-doped fluorotellurite glasses. Opt. Exp. **20**(4), 3803 (2012)
72. Rivera, V.A.G., El-Amraoui, M., Ledemi, Y., Messaddeq, Y., Marega Jr., E.: Expanding broadband emission in the near-IR via energy transfer between Er^{3+}–Tm^{3+} co-doped tellurite-glasses. J. Lumin. **145**, 787 (2014)
73. Deng, H., Yang, S., Xiao, S., Gong, H.M., Wang, Q.Q.: Controlled synthesis and upconverted avalanche luminescence of Cerium(III) and Neodymium(III) orthovanadate nanocrystals with high uniformity of size and shape. J. Am. Chem. Soc. **130**, 2032 (2008)
74. Ledemi, Y., Manzani, D., Sidney, J.L., Messaddeq, R.Y.: Multicolor up conversion emission and color tunability in Yb^{3+}/Tm^{3+}/Ho^{3+} triply doped heavy metal oxide glasses. Opt. Mat. **33**(12), 1916 (2011)
75. Rivera, V.A.G., Ledemi, Y., El-Amraoui, M., Messaddeq, Y., Marega, E. Jr., Green-to-Red Light Tuning by Up-Conversion Emission Via Energy Transfer in Er3+-Tm3+-Codoped Germanium-Tellurite Glasses. (2014). doi: 10.1016/j.jnoncrysol.2014.04.007
76. Haase, M., Schfer, H.: Upconverting nanoparticles. Angew. Chem. Int. Ed. **50**, 5808 (2011)
77. Zhao, J., Jin, D., Schartner, E.P., Lu, Y., Liu, Y., Vyagin, A.V., Zhang, L., Dawes, J.M., Xi, P., Piper, J.A., Goldys, E.M., Monro, T.M.: Single-nanocrystal sensitivity achieved by enhanced upconversion luminescence. Nat. Nanotechnol. **8**, 729 (2013)

78. Ledemi, Y., Trudel, A.A., Rivera, V.A.G., Chenu, S., Véron, E., Nunes, L.A.O., Allix, M., Messaddeq, Y.: White Light and Multicolor Emission Tuning in Triply Doped $Yb^{3+}/Tm^{3+}/Er^{3+}$ Novel Fluoro-Phosphate Transparent Glass-Ceramics. (2014). doi: 10.1039/C4TC00455H

79. Shalava, A., Richards, B.S., Green, M.A.: Luminescent layers for enhanced silicon solar cell performance: up-conversion. Sol. Energy Mater. Sol. Cells **91**, 829 (2007)

80. Wyatt, R.: Spectroscopy of rare earth doped fibres. Proc. SPIE **1171**, 54 (1990)

81. Kenyon, A.J.: Recent developments in rare-earth doped materials for optoelectronics. Prog. Quant. Electron. **26**, 225 (2002)

82. Van Uitert, L.G., Johnson, L.F.: Energy transfer between rare-earth ions. J. Chem. Phys. **44**, 3514 (1966)

83. Forster, T.: Zwischenmolekulare energiewanderung und fluoreszenz. Ann. Phys. **2**, 55 (1948)

84. Dexter, D.L.: A theory of sensitized luminescence in solids. J. Chem. Phys. **21**, 836 (1953)

85. Lakowicz, J.R.: Principles of Fluorescence Spectroscopy, 3rd edn. Springer Science + Business Media, New York, NY (2006)

86. Jorgensen, C.K., Judd, B.R.: Hypersensitive pseudoquadrupole transitions in lanthanides. Mol. Phys. **8**, 281 (1964)

87. Michael Reid, F., Richardson, F.S.: Electric dipole intensity parameters for lanthanide 4f → 4f transitions. J. Chem. Phys. **79**(12), 5735 (1983)

88. Malta, O.L., Luís, D.C.: Intensities of 4f-4f transitions in glass materials. Quim. Nova **26**(6), 889 (2003)

89. Rivera, V.A.G., Ledemi, Y., Osorio, S.P.A., Ferri, F.A., Messaddeq, Y., Nunes, L.A.O., Marega Jr., E.: Optical gain medium for plasmonic devices. Proc. SPIE **8621**, 86211J (2013)

90. Gopinath, A., Boriskina, S.V., Yerci, S., Li, R., Dal, N.L.: Enhancement of the 1.54 μm Er^{3+} emission from quasiperiodic plasmonic arrays. Appl. Phys. Lett. **96**, 071113 (2010)

91. Lakowicz, J.R.: Radiative decay engineering 5: metal-enhanced fluorescence and plasmon emission. Anal. Biochem. **337**, 171 (2005)

92. Tripathi, G., Rai, V.K., Rai, A., Rai, S.B.: Energy transfer between $Er^{3+}:Sm^{3+}$codoped TeO_2–Li_2O glass. Spectrochim. Acta. Part A **71**, 486 (2008)

93. Miroshnichenko, A.E., Flach, S., Kivshar, Y.S.: Fano resonances in nanoscale structures. Rev. Mod. Phys. **82**, 2257 (2010)

Chapter 3
Plasmonic Nanostructure Arrays Coupled with a Quantum Emitter

3.1 Introduction

In the previous chapter, we saw that a metal nanoparticle (NP) can generate localized surface plasmon resonance (LSPR), as a result of the resonance of free electrons on a curved surface. We also mentioned the main difference between surface plasmon polaritons (SPP) and LSPR, namely that SPPs are propagating waves, while LSPR is a nonpropagating excitation of free electrons in a metallic nanostructure coupled to an EM field. In this chapter, we will see that SPPs in restricted geometries (such as nanoantennas or nanocavities) can also generate LSPR. Nevertheless, the conditions for excitation are different. LSPR can be excited by direct application of light, whereas SPPs can be excited by matching the frequency and momentum of the excitation light and the SPPs. For example, under laser illumination, an antenna develops a strong dipole along the antenna. Charge oscillations inside each arm give rise to strong EM fields inside the feedgap of the antenna. When the antenna is resonant at optical frequencies, it can be called a resonant nanoantenna. In this case, with sufficiently close spacing with respect to a quantum emitter (QE) and nanoantenna, they can interact forming an electric dipole.

On the other hand, the extremely high localized fields in plasmonic nanocavities or plasmonic resonators, although they generally have low quality factors Q, are able to confine modes to sub-wavelength volumes V. Thus, plasmonic nanocavities become much more competitive with conventional optical resonators for applications where a resonance confined to a small volume is desirable. Hence, a nanocavity can profoundly change the number of EM modes and decay pathways available to a QE, increasing or decreasing its radiative decay rate. In both cases (nanoantenna and nanocavity), we can manipulate the local density of optical states (LDOS) of the QE, and increase the efficiency of light emission from the QEs.

© V.A.G. Rivera, O.B. Silva, Y. Ledemi, Y. Messaddeq, and E. Marega Jr. 2015
V.A.G. Rivera et al., *Collective Plasmon-Modes in Gain Media*, SpringerBriefs
in Physics, DOI 10.1007/978-3-319-09525-7_3

In this scenario, a hybrid resonance mode allows us to combine the best of both worlds: the multiple-frequency and high-Q properties of photonic components with the nanoscale dimensions of plasmons, leading to nanophotonics with a wealth of "new" effects based on quantum plasmonics.

3.2 Extraordinary Optical Transmission: Optical Nanoantennas

With exciting advances in nanotechnology, there exists a renewed interest in exploiting the dielectric properties of materials, in order to create photonic materials able to overcome the limits of optical diffraction. This limit was achieved in 1998 by Ebbesen et al. [1], who verified the transmission of light through nanoholes (of size d) made in a metal film (of thickness D) via SPP tunneling, since the walls are metal (and perfect conductors) and can therefore propagate the incident light in the form of SPPs through the nanohole. The transmitted intensity of this SPP propagation has approximately exponential dependence on the thickness D [2]. This effect is called extraordinary optical transmission (EOT). By using scanning near-field optical microscopy (SNOM), it can be shown that the nanohole acts as a scattering center for SPPs [3, 4]. On the other hand, by using perfect conductors for the walls, we cannot neglect the penetration of the EM fields through the films, since, in the optical regime, EM fields penetrate the metal up to a distance called the skin depth of the metal. For example, in a planar geometry, the skin depth of a metal is [5]:

$$\delta(\omega) = \frac{\sqrt{2}c}{\omega} \left\{ \sqrt{\mathrm{Re}[\varepsilon_m(\omega)]^2 + \mathrm{Im}[\varepsilon_m(\omega)]^2} - \mathrm{Re}[\varepsilon_m(\omega)] \right\}^{-1/2}. \qquad (3.1)$$

We can see that the skin depth is also a function of ω. In consequence, the coupling between the SPPs at the two sides of the film results in hybridization between the two modes. Hence, we have two coupled modes ω_+ and ω_- of the thin film [6]:

$$\omega_\pm(p,D) = \frac{\omega_{spp}}{\sqrt{2}} \left(1 \pm e^{-pD}\right)^{1/2}. \qquad (3.2)$$

The energy splitting between the two modes depends on the film thickness D and the parallel momentum p. At the thick-film limit, the coupling between the two SPPs of the film becomes weak.

For a plane wave with normal incidence, wavelength λ, intensity I_0 over the area of the nanoaperture, and transverse-electric (TE) or transverse-magnetic (TM) polarization, a nanoaperture can be described as a magnetic dipole located in the plane of the nanohole [7, 8]. Therefore, the transmission coefficient is given by Bethe's theory:

$$T_r = \frac{64}{27\pi^2}\left(kd'\right)^4 \propto \left(\frac{d'}{\lambda}\right)^4, \tag{3.3}$$

where d' is the hole size. The proportionality $T_r \propto \lambda^{-4}$ is in agreement with Rayleigh's theory of scattering by small objects. However, this is smaller when compared with Gustav Kirchhoff scalar diffraction ($T_r \propto \lambda^{-2}$) [9]. On the other hand, the results show that Bethe's theory, Eq. (3.3), is insufficient for describing a real metal system. This is because there is another SPP mode existing in the nanoholes, besides the SPP mode excited on the film surface around the nanohole [10–12]. This additional SPP mode, originating from the coupling of light with electrons on the nanohole walls, is mostly localized in the dielectric core and travels (propagating or evanescent) along the nanohole axis. It has been recently shown both theoretically and experimentally that the transmission peak of the sub-wavelength hole can be attributed to a Fabry–Perot resonance owing to multiple reflections of the fundamental cavity mode [13, 17].

When nanoholes form a two-dimensional (2D) array, the transmission of light through these structures can have maxima (as a periodic array of nanoholes transmits more light than a large macroscopic hole with the same total area as the nanoholes) or minima at some particular wavelengths [14]. The maxima/minima (resonance enhancement/suppression, respectively) of the transmission peaks appearing in the transmission spectra can be explained by the dispersion relation of SPPs modes propagating on the metal surface. An EOT enhancement is due to phase matching of the incident radiation with SPPs; the same process should occur for a single nanohole surrounded by a regular array of opaque surface corrugations.

The resonance suppression of transmission has been experimentally observed as asymmetric line shapes that agree with Fano's theory [15]. Therefore, it is clear that there is a close connection between EOT and the excitation of SPPs.

A single slit is another interesting structure, as it combines the compactness of a single defect with the directionality of a periodic nanostructure, since the long axis imposes directionality on the transmitted light beam [2, 16]. Furthermore, EOT from a single slit is noticeably affected by the film thickness, and increases linearly with increasing slit width for a fixed film thickness [17]. These effects can be improved when a slit is surrounded by a regular array of parallel grooves.

As these 2D periodic nanostructures exhibit optical properties similar to 2D photonic crystals, they can be used to guide SPP waves in linear or bent structures [18], enhance EM field confinement within nanoholes [19], and achieve nonlinear optical effects with smaller light intensities than in conventional photonic crystals [20–22]. A periodic array (a plasmonic crystal) can convert out-of-plane multi-wavelength input into in-plane SPP output with different light frequencies propagating in different directions on a smooth or corrugated surface. Additionally, the size and shape of the nanoholes and their periodicity can then be optimized to realize the ideal SPP dispersion required for applications [23, 24]. We should also mention the sensitivity of the coupling between the SPP and the polarization of the incident light to the longitudinal nature of the SPP field, and we note that the excitation field should have an electric field perpendicular to the surface or parallel to the direction of SPP propagation [25, 26].

Concentric circles (slits or grooves) surrounding a nanoaperture center (called a bull's eye structure) can also be used to accomplish phase matching of the incident radiation to the SPPs. Here, the enhancement of transmission is compared to the values calculated from Eq. (3.3). The height, h, of the undulations responsible for coupling determines the efficiency of SPP coupling, and therefore the magnitude of the transmission [27, 28].

By altering the parameters of the ring-grooves (e.g., h, ring-groove spacing, and number of ring-grooves), the diffracted beams can be manipulated and the intensity of the SPP (I_{spp}) in the plasmonic planar lens can be enhanced [28]. In this manner, the spacing between the rings determines the phase mismatch between the waves generated from the inner and outer rings. By tuning to the λ_{spp}, the periodicity provides momentum matching along the ring-grooves, ensuring excitation of resonant SPPs at the foci. The periodic corrugations also become active coupling elements that depend on the ring-groove spacing. Therefore, this focusing ensures a high confinement in the center of the "bull-eye" (BE) structure, which enhances the EOT in the nanohole incrementally with the ring-grooves. The total transmittance intensity through the hole can be expressed as [29]:

$$I_{\text{total}} = \sum_{i=1}^{N} CI_{\text{spp}} I_0 \frac{4}{r_i \lambda_{\text{spp}}} \exp\left\{ \frac{-|r - r_i|}{L_{\text{spp}}} \right\} \propto \left(\frac{d}{\lambda} \right)^4, \tag{3.4}$$

where r is the radius of the central hole, r_i is the radius of each ring, i is the number of the ith ring-groove, and C is the coupling efficiency of the groove [30]. Furthermore, when SPP waves propagate across a single ring, the transmitted intensity will be modified by the wave that gains a phase-shift of Φ [31]. Consequently, each ring-groove of the plasmonic lens re-illuminates the BE via the SPPs.

In this manner, metallic nanostructures sustaining surface plasmons can change the excitation and emission properties of locally excited quantum systems such as rare-earth ions (REI).

In this context, the big advances in nanofabrication and EM simulation techniques allow us to realize a wide variety of plasmonic structures that exploit EOT effects. Thus, noble metal nanostructures may possess exactly the right combination of electronic and optical properties to tackle the issues of the optical integration and realize the dream of numerous research groups and industries of significantly faster processing speeds at the nanoscale. Consequently, EOT has enabled the fabrication of several devices with extraordinary capabilities, including high-resolution optical spectroscopy and microscopy [32, 33], nanolithography [34], high-capacity data storage [35], enhanced nonlinear phenomena [36], enhanced light emission and lasing [37, 38], high-efficiency and high-speed photodetection [39], and others.

3.3 Localized Surface Plasmons in Metal Nanostructure Arrays

SPPs constitute a coupling between the EM radiation and the charge density at the interface of two media. This coupling arises from matching the phase of electron oscillations with the incoming field. The propagation of an SPP at the interface is completely known when we solve Maxwell's equations for this region, applying the pertinent boundary conditions for the fields. Consequently, we obtain the dispersion relation for the SPP that relates the component of the wave vector parallel to the interface with the radiation frequency. The basic assumption to consider is that both media are linear, homogenous, isotropic, and nonmagnetic ($\mu = 1$) materials, with the dispersion properties defined by their respective permittivities (or dielectric functions) that have already been analyzed, especially for metals, as discussed in Chap. 1. For simplicity, we adopt as our model the planar interface between two media, with permittivities $\varepsilon_m(\omega)$ and $\varepsilon_d(\omega)$, at the plane $z = 0$. This structure is depicted in Fig. 1.3a.

The study of plasmon propagation at this interface begins with Maxwell's equations:

$$\nabla \cdot \boldsymbol{D} = 0, \tag{3.5}$$

$$\nabla \cdot \boldsymbol{B} = 0, \tag{3.6}$$

$$\nabla \times \boldsymbol{E} = -\frac{\partial \boldsymbol{B}}{\partial t}, \tag{3.7}$$

and

$$\nabla \times \boldsymbol{H} = \frac{\partial \boldsymbol{D}}{\partial t}. \tag{3.8}$$

Both materials are source-free, i.e., there is no net charge ρ and no current density $\boldsymbol{J}$. By applying the constitutive relations $\boldsymbol{D} = \varepsilon_0 \varepsilon \boldsymbol{E}$ and $\boldsymbol{B} = \mu_0 \boldsymbol{H}$ in Eqs. (3.7) and (3.8), we obtain the following expressions:

$$\nabla \times \boldsymbol{E} = -\mu_0 \frac{\partial \boldsymbol{H}}{\partial t} \tag{3.9}$$

and

$$\nabla \times \boldsymbol{H} = \varepsilon_0 \varepsilon_m \frac{\partial \boldsymbol{E}}{\partial t}, \tag{3.10}$$

where ε_0 corresponds to the electric permittivity for the vacuum and μ_0 is the magnetic permeability for the vacuum. It is possible to obtain the general wave equation in terms of the electric or magnetic field. For convenience, we will write it as a function of the electric field. Applying the curl operator to both sides of Eq. (3.9) gives:

$$\nabla \times (\nabla \times E) = -\mu_0 \varepsilon_0 \varepsilon_m \frac{\partial^2 E}{\partial t^2}. \tag{3.11}$$

The right-hand side of this equation is achieved with the result from Eq. (3.10). A further development of this equation leads to:

$$\nabla(\nabla \cdot E) - \nabla^2 E = -\mu_0 \varepsilon_0 \varepsilon_m \frac{\partial^2 E}{\partial t^2}. \tag{3.12}$$

A particular feature of the first element from the right-hand side of the above equation above deserves our attention. From vector identities:

$$\nabla \cdot (\varepsilon E) = E \cdot \nabla \varepsilon + \varepsilon \nabla \cdot E \rightarrow \nabla \cdot E = -\frac{E \cdot \nabla \varepsilon}{\varepsilon}. \tag{3.13}$$

Since there is no charge source $\nabla \cdot D = \nabla \cdot (\varepsilon_0 \varepsilon_m E) = 0$, it is possible to isolate the term from the last equation. Applying Eq. (3.13) on Eq. (3.12) gives:

$$\nabla \left(-\frac{E \cdot \nabla \varepsilon}{\varepsilon} \right) - \nabla^2 E = -\mu_0 \varepsilon_0 \varepsilon_m \frac{\partial^2 E}{\partial t^2}. \tag{3.14}$$

In general, the permittivity has a spatial and spectral dependence, i.e., $\varepsilon = \varepsilon(r, \omega)$. Nevertheless, for distances smaller than the wavelength, the spatial dependence becomes irrelevant, $\nabla \varepsilon \approx 0$. Therefore, the gradient of the permittivity is neglected. The remaining terms of the last expression define the wave equation:

$$\nabla^2 E - \frac{\varepsilon_m}{c^2} \frac{\partial^2 E}{\partial t^2} = 0. \tag{3.15}$$

We used the relation $\mu_0 \varepsilon_0 = \frac{1}{c^2}$.

An important approximation to solve this equation is to consider the field as exhibiting harmonic time dependence, i.e., $E(r, t) = E(r)e^{-i\omega t}$. Consequently, it is possible to write the time operator as $\frac{\partial^2}{\partial t^2} \propto -\omega^2$, and the wave equation is rewritten as:

$$\nabla^2 E + \varepsilon_m k_0^2 E = 0. \tag{3.16}$$

This is called the Helmholtz equation. The solutions of the Helmholtz equation define the spatial profile of the field's components or modes. Furthermore, the application of boundary conditions for these components at the interface leads to the dispersion relation. Our goal is to look for the dispersion relation of the surface plasmons. Establishing these two assumptions is essential to solving this problem: (1) the incident EM wave displays TM polarization (also known as a p-wave), otherwise there would be no formation of a surface plasmon, and (2) since we are interested only in solutions that describe the propagation near the interface, there will be an evanescent wave in the perpendicular direction of the interface, i.e., for regions far away from the limit between two media there is no dependence on the z-component.

For TM polarization, the nonvanishing components of the electric field are E_x and E_z. In this manner, we can defining $\boldsymbol{E_m}$ as the electric field on the ar-metal interface, and $\boldsymbol{E_d}$ as the electric field on the metal-dielectric interface (Fig. 1.3a), as follows:

$$\boldsymbol{E}_m = E_{mx}e^{ik_{mx}x-i\omega t}e^{ik_{mz}z}\hat{x} + E_{mz}e^{ik_{mx}x-i\omega t}e^{ik_{mz}z}\hat{z} \tag{3.17}$$

and

$$\boldsymbol{E}_d = E_{dx}e^{ik_{dx}x-i\omega t}e^{ik_{dz}z}\hat{x} + E_{dz}e^{ik_{dx}x-i\omega t}e^{ik_{dz}z}\hat{z}. \tag{3.18}$$

The wave vector component parallel to the interface is conserved ($k_{1x}=k_{2x}=k_x$). Therefore, it is possible to rewrite the above equations in a simpler form:

$$\boldsymbol{E}_m = (E_{mx}\hat{x} + E_{mz}\hat{z})e^{ik_x x-i\omega t}e^{ik_{mz}z} \tag{3.19}$$

and

$$\boldsymbol{E}_d = (E_{dx}\hat{x} + E_{dz}\hat{z})e^{ik_x x-i\omega t}e^{ik_{dz}z}. \tag{3.20}$$

Examining the components of the wave vector is important for our purpose. The first component is related to the modulus of the wave vector in each medium:

$$k_x^2 + k_{j,z}^2 = \varepsilon_j k_0^2 \quad j = m, d, \tag{3.21}$$

where the index j denotes the medium that the field passes through. The second is a consequence of Maxwell's equations. Equation (3.5) involves the dielectric displacement vector, as $\nabla \cdot \boldsymbol{D} = 0$, and, since the spatial dependence for the fields is of the form $e^{i\boldsymbol{k}\cdot\boldsymbol{r}}$, it is reasonable rewrite the divergence operator as $\nabla \cdot \propto i\boldsymbol{k}$. With this information, we obtain $\boldsymbol{k} \cdot \boldsymbol{D} = 0$. Additionally, from the constitutive relation $\boldsymbol{D} = \varepsilon_0 \varepsilon_j \boldsymbol{E}$, it is possible to establish other useful expression for the components of $\boldsymbol{k}$:

$$k_x E_{j,x} + k_{j,z}E_{j,z} = 0. \tag{3.22}$$

Furthermore, the boundary conditions provide the necessary information about the components of the electric field. The first condition establishes the continuity of the parallel components of the electric field at the interface:

$$\hat{n} \times E = 0 \rightarrow E_{1,x} - E_{2,x} = 0. \tag{3.23}$$

The second condition is about the continuity of normal components of the dielectric displacement D:

$$\hat{n} \cdot D = 0 \rightarrow \varepsilon_1 E_{1,z} - \varepsilon_2 E_{2,z} = 0. \tag{3.24}$$

For both conditions, $\hat{n}$ corresponds to the normal unitary vector at the interface. Equations (3.22), (3.23), and (3.24) form a homogenous system for the four unknown electric field components, as follows:

$$k_x E_{m,x} + k_{m,z} E_{m,z} = 0$$
$$k_x E_{d,x} + k_{d,z} E_{d,z} = 0$$
$$E_{m,x} - E_{d,x} = 0$$
$$\varepsilon_m E_{m,z} - \varepsilon_d E_{d,z} = 0.$$

It is possible rewrite the initial system in matrix notation:

$$\begin{pmatrix} k_x & 0 & k_{m,z} & 0 \\ 0 & k_x & 0 & k_{d,z} \\ 1 & -1 & 0 & 0 \\ 0 & 0 & \varepsilon_m & -\varepsilon_d \end{pmatrix} \begin{pmatrix} E_{m,x} \\ E_{d,x} \\ E_{m,z} \\ E_{d,z} \end{pmatrix} = \begin{pmatrix} 0 \\ 0 \\ 0 \\ 0 \end{pmatrix}.$$

For solutions to exist, the associated determinant must vanish. After algebraic manipulation, the calculation of the determinant under the latter condition leads to:

$$k_x(\varepsilon_m k_{d,z} - \varepsilon_d k_{m,z}) = 0. \tag{3.25}$$

The equation above contains valuable information. The system admits only two sets of solutions. One of them is the trivial solution $k_x = 0$, i.e., there is no propagation along the surface between the two media. Such a case is not valid in the description of surface waves. Otherwise, the second set indicates that the term inside the parentheses from the equation above is null. This condition constitutes the physical meaning of the propagation of the SPP. From Eq. (3.21), we can rewrite the normal components of the wave vector in terms of k_x:

$$k_{j,z} = \sqrt{\varepsilon_j k_0^2 - k_x^2} \quad j = m, d. \tag{3.26}$$

Applying this result to the condition for the valid existence of a solution gives:

$$\varepsilon_m \sqrt{\varepsilon_d k_0^2 - k_x^2} - \varepsilon_d \sqrt{\varepsilon_m k_0^2 - k_x^2} = 0. \tag{3.27}$$

After some algebraic manipulation, we obtain the expression for the parallel component of the wave vector as a function of the respective permittivity of the two media:

$$k_x = k_0 \sqrt{\frac{\varepsilon_m \varepsilon_d}{\varepsilon_m + \varepsilon_d}}. \tag{3.28}$$

The preceding expression is also known as the dispersion relation for the surface waves and k_x denotes the propagation constant of the surface waves. Additionally, we are able to write the corresponding dispersion relation for the normal component of the wave vector, applying the expression for k_x in Eq. (3.26) to yield:

$$k_{j,z}^2 = \frac{\varepsilon_j^2}{\varepsilon_m + \varepsilon_d} k_0^2. \tag{3.29}$$

Since we have obtained the dispersion relations, we are able to define the conditions for the propagation of an SPP. To describe surface waves, the propagation constant k_x must be a real quantity. Such a condition is completely fulfilled if the numerator and denominator of k_x are both positive or both negative. Furthermore, as we have discussed before, since we are looking for solutions near to the interface, there is an evanescent wave in the normal direction of the interface, i.e., the wave in the z-direction decays exponentially. To fulfill this condition, the normal component of the wave vector must be purely imaginary. By analyzing the dispersion of Eq. (3.29), $k_{j,z}$ will be imaginary if only and only if the sum in the denominator is negative. Consequently, from the first condition, the product of two dielectric functions is also negative. In summary, the two conditions for the propagation of SPP waves that need to be fulfilled are:

$$\varepsilon_m(\omega).\varepsilon_d(\omega) < 0 \tag{3.30}$$

and

$$\varepsilon_m(\omega) + \varepsilon_d(\omega) < 0. \tag{3.31}$$

One basic group of materials that exhibit dispersive properties that fulfill the above conditions are the metals. They have dielectric functions with large negative real parts, as the graphs for the dielectric functions in Fig. 1.1, Chap. 1, have shown. As an example, we suppose that medium 1 is gold, while medium 2 is a material that has a permittivity with a small positive real part, in which the absolute value is lower than that of gold. Dielectric materials, like air, glass, or water, for instance, possess such properties. In general, the permittivity for a dielectric material is purely real. The combination of these two sorts of media allows the propagation of SPP modes. It is well known in the literature that SPP waves arise from a metal–dielectric interface. In addition to these conditions, the excitation of SPP waves also depends on the features of the wave vector of the incident radiation. In order to comprehend the excitation of SPPs and its relationship to the wave vector, it is necessary to analyze the dispersion of SPP modes, i.e., the dependency of frequency ω on the parallel component of the wave vector k_x.

As we have seen, a more reliable method to describe the dielectric function of a metal is to consider the interband transitions resulting from bound electrons. However, for a simple comparison, we will also include the Drude model in the analysis of the dispersion relation, because, although it is a limited model, it presents similar features to the interband transitions. Figure 3.2 shows the dispersion relation for SPPs.

Since the dielectric medium adopted to plot these dispersion relations was air, ε_d is a real quantity. Furthermore, since the x-component of the wave vector is proportional to $\frac{1}{\lambda}$ for small values of k_x (at the infrared region of the spectrum), the Drude model and interband transitions effects describe the same features; however, as k_x increases (in the visible and UV regions), the two models present differences. Such behavior is in clear agreement with the dielectric function for metals, as shown in Figs. 1.1 and 1.2. Both models have two branches for their dispersion relations: one top branch for high energies, and one bottom branch for low energies. Surface waves are not described by the high-energy branch, since, in this case, the component of the wave vector normal to the metal–dielectric interface k_z has a real part, in contrast to the conditions that have established previously. Nevertheless, the low-energy branch describes exactly the surface waves.

In order to describe completely the properties of SPPs, it is necessary consider ohmic losses of the metal's electrons. For this reason, the dielectric function ε_m from Eq. (3.28) has an imaginary part, ($\varepsilon_m = \varepsilon'_m + i\varepsilon''_m$). Consequently, the wave vector k_x is complex; its real and imaginary parts possess important physical meanings. The real part defines the SPP wavelength λ_{SPP}:

$$\lambda_{SPP} = \frac{2\pi}{\text{Re}[k_x]}, \tag{3.32}$$

which constitutes the wavelength of the SPP waves at the metal–dielectric interface.

Since metals possess a dielectric function with the real part greater than the imaginary part ($|\text{Re}[\varepsilon_m]| \gg |\text{Im}[\varepsilon_m]|$), λ_{SPP} can be approximated by the following expression:

$$\lambda_{SPP} \approx \sqrt{\frac{\text{Re}[\varepsilon_m] + \varepsilon_d}{\text{Re}[\varepsilon_m]\varepsilon_d}}\lambda_0. \tag{3.33}$$

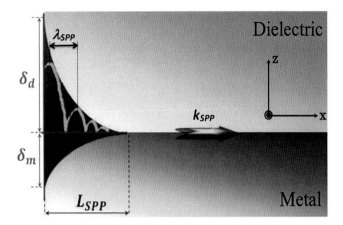

Fig. 3.1 Illustration of SPP coupling at an interface between a metal (medium 1) and a dielectric (medium 2), and its parameters

where λ_0 is the wavelength in free space. We have applied the results from Eq. (3.28) to obtain the relation above. The corresponding wave vector $k_{SPP} = \frac{2\pi}{\lambda_{SPP}}$ denotes the parallel component of an incident radiation wave vector responsible for the coupling (or excitation) of the SPP.

Otherwise, the imaginary part is related to the propagation length L_{SPP}. This quantity refers to the damping of the electric field amplitude resulting from the loss of electron, and it is defined by:

$$L_{SPP} = \frac{1}{2\text{Im}(k_x)}. \tag{3.34}$$

The propagation length can also be interpreted as the length that the SPP waves propagate along the interface as the amplitude of the electric field decays by $1/e$ from its initial value. For typical noble metals in the visible regime, L_{SPP} is about a dozen micrometers. A similar consideration for the normal component of the wave vector to the interface k_z defines the length related to the exponential decay of radiation away from interface, also known as the skin depth $\delta = \frac{1}{k_z}$. Since losses in metals are more significant, the penetration of the electric field into this medium is shorter compared to that in the dielectric, as shown in Fig. 3.1.

Another relevant characteristic of the SPP is that the wave vector k_x of light at the interface is larger compared to that in free space (the "light line" on the dispersion relation curve). A larger wave vector denotes an increased momentum for the radiation at the interface for the same photon energy $\hbar\omega$ in free space. This increase is a consequence of the strong coupling between the radiation and the surface charges on the metal. Therefore, in order to excite the SPP at a planar metal–dielectric interface, the radiation in free space alone is insufficient to couple the plasmons; it is necessary for the parallel component of the incident wave vector

Table 3.1 Wavelengths in free space and in the metallic surface

Wavelengths in free space versus the wavelength of the SPP modes		
Metal	λ_0 (nm)	λ_{SPP} (nm)
Gold	632.8	373.1
Silver	488	277.4

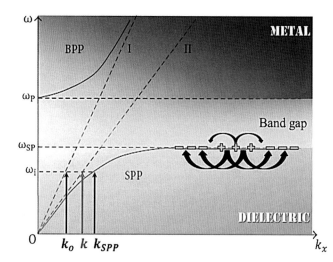

Fig. 3.2 The dispersion relation for SPP modes (lower branch). There are two "light lines" (the *straight lines*) depicted: *line I* is related to the propagation of light in air and *line II* is associated with propagation into glass. For the same energy with frequency ω_i, air and glass possess different x-components for the wave vector. The SPP excitation is clearly enhanced for glass, since the wave vector inside it (the *green k*) is closer to k_{spp} than the wave vector inside air, k_0. The *upper branch* of the dispersion relation is related to bulk plasmons, where $\omega_{spp} = \omega_p(1 - \text{Im}[\varepsilon])^{0.5}$

to match with k_x. Table 3.1 compares the wavelengths in free space and in typical noble metals that have been used in plasmonics. The decreasing wavelength denotes an increase in the wave vector.

$$\varepsilon_{Au} = -11.740 + i1.261; \; \varepsilon_{Ag} = -9.122 + i0.302; \; \varepsilon_{glass} = 2.3104$$

In order to overcome the problem of SPP excitation, the simplest solution is to replace the air with another dielectric medium that possesses a higher refractive index. Since the dispersion relation in free space is expressed by $\omega = ck_x$, when the dielectric medium is replaced by glass, for instance, the SPP excitation is enhanced because, for this new situation, the dispersion relation in glass is expressed by $\omega = \frac{c}{n}k_x$. Because $n > 1$, the "light line" for the glass approaches the dispersion curve of the SPP modes. Therefore, the choice of a dielectric that possesses a high refractive index is critical for the enhancement of SPP coupling (Fig. 3.2).

3.3.1 Resonance Modes and Tuning

In the last decade, plasmonic nanostructures have been extensively investigated, revealing amazing effects and fascinating phenomena, such as EOT, giant field enhancement, SPP nanoguides, and recently metamaterials have emerged, which are often based on plasmonic nanostructures [40]. One of the most exciting features is the ability to tune the resonant modes in the plane of the nanostructure, mainly owing to the propagating SPP (λ_{SPP}) interacting with the various nanostructures, generating LSPR with the same λ_{SPP}. These traveling waves are continually interacting on the surface (under continuous excitation), i.e., constructive and destructive interference, and depend on the periodic structures or surface corrugations. We recall that we can also have hybridization of the SPP on both sides of the metal surface, see Eq. (3.2), and that the transmission of incident light through the metal film can create induced surface plasmons (ISP), as in Eq. (3.3). However, both cases can be neglected at the thick-film limit. For the one-dimensional (1D) case of periodic slits in metal films, three mechanisms were identified: (1) SPP (2) waveguide modes, and (3) optical cavities in slits [17, 41, 42]. Andrewartha et al. [43] studied gratings in perfect conductors and revealed that modes exist inside the grooves. Such modes cause a strong redistribution of energy into the different diffracted orders, which depend on the groove width, groove depth, and wavelength. These resonance cavities can explain the anomalies in the orders of reflection from the gratings for both TE and TM polarization.

Within these approximations, the resonance modes and tuning of the transmission properties of a periodic nanohole array (2D) can be explained in terms of the Bloch EM modes in each spatial region [23, 44, 45] determined for a unit cell of the nanostructure. Each nanohole is considered to be an optical cavity providing feedback of the resonance mode. This nanostructure is conventionally illuminated with a light beam (Fig. 3.3, with the configuration most widely used in experiments). Then SPP propagation has the same effects as electrons propagating through a crystal lattice or photons propagating through a medium with a periodically modulated refractive index (photonic crystals), i.e., we have SPP eigenmodes (Bloch waves) propagating on a periodic surface, which can excite certain frequencies in specific directions, where the diffraction provides conservation of the wave vector:

$$k_{spp} = \frac{\omega}{c} n_{eff} \sin\theta \, \vec{u}_{xy} \delta_p \pm i \frac{2\pi}{a_0} \vec{u}_x \pm j \frac{2\pi}{a_0} \vec{u}_y. \tag{3.35}$$

These SPPs correspond to standing SPP Bloch waves on a periodic surface. Here, k_{spp} is the SPP Bloch wave vector on the periodic nanostructure, n_{eff} is the effective refractive index of the air–metal interface, θ is the angle of incidence, $\vec{u}_{xy}$ is the unit vector in the plane of the film in the direction of projection of the incident light wave vector, $\delta_p = 0$ or 1 for p- or s-polarized incident light (relative to the

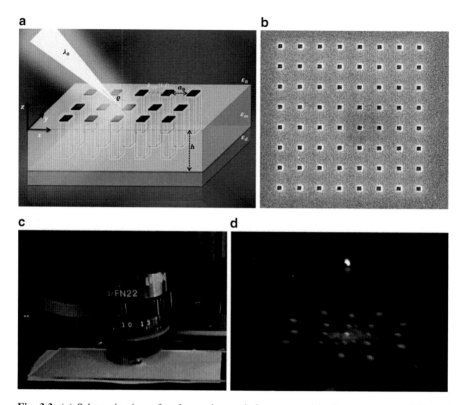

Fig. 3.3 (a) Schematic view of a plasmonic nanohole array used in the study: a metal film of thickness h on a dielectric substrate with the physical parameters of the system. (b) Scanning electron microscope (SEM) images of a nanostructure array of approximately $10 \times 10 \ \mu m^2$ with 8×8 *squares* milled in *gold film*. (c) Photo of the experimental setup used for microluminescence measurements of a plasmonic array formed on Er^{3+}-doped tellurite glass; the sample was excited with a diode laser at 405 nm. (c) Photo of the sample shown in (b) with a *black* background. We can see the diffraction in the *white paper*, produced by EOT of the *blue light* through the *square* nanohole. The *green light* is the fluorescence produced by the erbium ions within the glass

sample surface), respectively, $\vec{u}_x$ and $\vec{u}_y$ are the unit reciprocal lattice vectors of the periodic structure, a_0 is the periodicity (same in the x- and y-directions), and i and j are integer numbers corresponding to different directions in the SPP Brillouin zone and determining the direction of the SPP on a metallic surface.

Therefore, in the vicinity of a periodically nanostructured surface, both the SPP and light coexist, with the SPP propagating along the interface in $z = 0$ (Fig. 3.3). From this, the EM field of the SPP has both transverse and longitudinal components (x-axis), which can be described by:

$$E_{spp} = (E_x, 0, E_z)e^{ik_x x - i\omega t} + c.c \tag{3.36}$$

and

$$H_{\text{spp}} = (0, H_y, 0) e^{ik_x x - i\omega t} + c.c. \tag{3.37}$$

Two-dimensional finite periodic nanostructures can match the Φ for LSPR with the Bragg resonance of the grating. Thus, the optical response for a normal incidence can be a maximum or a minimum in both cases, depending on the wavelength of LSPR (λ_{spp}) and the symmetry of the nanohole array. For example, for normal incidence in a square lattice, we have $\theta = 0$ and $\left| \vec{u}_x \right| = \left| \vec{u}_y \right| = 2\pi/a_0$, and the transmission maxima can be estimated as [1]:

$$\lambda_{\text{spp}}(i,j) = \frac{a_0}{\sqrt{i^2 + j^2}} \text{Re}\left[\sqrt{\frac{\varepsilon_d \varepsilon_m}{\varepsilon_d + \varepsilon_m}} \right]. \tag{3.38}$$

The transmission minima correspond to a diffraction phenomenon known as Wood's anomaly [46]. This occurs when an order of diffraction becomes tangent to the plane of the grating. Then the intensity of the light is redistributed among the remaining orders and is given by:

$$\lambda_{\text{Wood}}(i,j) = \frac{a_0}{\sqrt{i^2 + j^2}} \sqrt{\varepsilon_d}. \tag{3.39}$$

On the other hand, for normal incidence in a hexagonal lattice, we have $\theta = 30°$ and $\left| \vec{u}_x \right| = \left| \vec{u}_y \right| = 4\pi/\sqrt{3}a_0$, and the transmission maxima can be estimated as:

$$\lambda_{\text{spp}}(i,j) = \frac{a_0}{\sqrt{\frac{4}{3}(i^2 + i \cdot j + j^2)}} \text{Re}\left[\sqrt{\frac{\varepsilon_d \varepsilon_m}{\varepsilon_d + \varepsilon_m}} \right]. \tag{3.40}$$

Because Eqs. (3.38), (3.39), and (3.40) do not take into account the presence of holes (their size or geometry) and the associated scattering losses, this neglects the interference that gives rise to a resonance shift. As a consequence, this predicts the peak positions at wavelengths slightly shorter than those observed experimentally. Other effects include the Rayleigh anomaly, when the light is diffracted to move parallel to the grating surface, and Fano resonance, which occurs when a discrete state (Bloch waves) and a continuum state (direct nanohole transmission) interact with one another.

Figure 3.4 shows theoretical and experimental results for the transmission spectra of two periodic arrays. The minima adjacent to each maximum in the transmission spectra of the 2D array are related to Eqs. (3.39) and (3.40), respectively, as shown in Fig. 3.4a. The red dotted line shows the EOT from a single square hole. Figure 3.4b shows the experimental results from ref. [48].

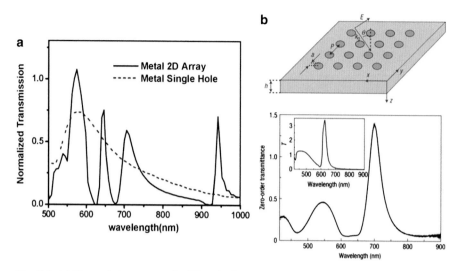

Fig. 3.4 (**a**) Transmission spectra for 2D periodic *square arrays* of $d = 200$ nm nanoholes with *square* lattice spacings of $a_0 = 600$ nm in a 100-nm-thick metal film. The figure extracted from ref. [47]. Reprinted under United States Copyright law (Optical Society American). (**b**) Transmission spectra for 2D periodic (*hole*) arrays. This figure extracted from ref. [48]. Reprinted with permission of copyright Nature publishing group

3.4   Propagation, Guiding, and Localization of Resonance Modes

Light can couple with SPPs if the surface in which the SPPs reside displays a definite nanostructure that satisfies conservation of energy and momentum. From this basic concept, we can have plasmonic signal generation, transport, modulation, and so forth, all integrated in a plasmonic circuit chip.

A plasmonic waveguide is one of the most important components for the realization of a plasmonic chip, in which the optical signal can be sent from one section to another at the nanoscale without the need to convert the light to an electric signal. In the same way that planar waveguides can transport light in one dimension, a planar metal dielectric interface can confine the light in the form of an SPP along this plane, also in one dimension. In the literature, different configurations of waveguide have been proposed and investigated, such as MDM, DMD, nanowires, NPs arrays, and others. However, all these waveguides are based on the existence and propagation of SPP modes at the interface between regular and negative dielectrics.

For a 1D plasmonic waveguide magnetic dipole, the theory of optical waveguides is based on solving Maxwell's equations in a linear isotropic media. Since nonmagnetic materials are commonly used for an optical waveguide, the magnetic permeability here is μ_0, and then the Maxwell's equations can be reduced to:

$$\Delta E(r, t) - \varepsilon_0 \varepsilon(r) \mu_0 \frac{\partial^2 E(r, t)}{\partial t^2} = \nabla(E(r, t) \cdot \nabla \ln\{\varepsilon_0 \varepsilon(r)\}) \tag{3.41}$$

and

$$\Delta H(r, t) - \varepsilon_0 \varepsilon(r) \mu_0 \frac{\partial^2 H(r, t)}{\partial t^2} = (\nabla \times H) \times (\nabla \ln\{\varepsilon_0 \varepsilon(r)\}). \tag{3.42}$$

The fields can be expressed as:

$$\begin{aligned} E &= e(x, y)\exp\{i(\beta z - \omega t)\}, \\ e(x, y) &= e_t(x, y) + e_z(x, y)z_0, \\ H &= h(x, y)\exp\{i(\beta z - \omega t)\} \end{aligned} \tag{3.43}$$

and

$$h(x, y) = h_t(x, y) + h_z(x, y)z_0, \tag{3.44}$$

where β denotes the propagation constant of a mode and the subscript t the transverse component of the field vectors. These equations contain all the information necessary to determine the modal fields and propagation constants of the modes guided for the waveguide. The modal fields are bounded in the transversal and z-directions, and decay fast at large distances from the waveguide (as evanescent modes). The propagation constant is related to the effective modal index n_{eff} and the modal attenuation b, as follows:

$$\mathrm{Re}[\beta] = \frac{\omega}{c} n_{\mathrm{eff}} \tag{3.45}$$

and

$$\mathrm{Im}[\beta] = b\frac{\ln 10}{0.2}, \tag{3.46}$$

where the attenuation is in dB cm^{-1} if β is given in m^{-1}.

If $\mathrm{Re}[\varepsilon_m(\omega)] < 0$ and $|\mathrm{Re}[\varepsilon_m(\omega)]| > |\mathrm{Im}[\varepsilon_m(\omega)]|$, then the complex propagation constant of the SPP can be expressed as:

$$\beta = \beta_R + i\beta_I = \frac{\omega}{c}\sqrt{\frac{\varepsilon_d \mathrm{Re}[\varepsilon_m]}{\varepsilon_d + \mathrm{Re}[\varepsilon_m]}} + i\frac{\omega \mathrm{Im}[\varepsilon_m]}{2c\mathrm{Re}[\varepsilon_m]^2}\sqrt{\left(\frac{\varepsilon_d \mathrm{Re}[\varepsilon_m]}{\varepsilon_d + \mathrm{Re}[\varepsilon_m]}\right)^3}, \tag{3.47}$$

where β_R and β_I are the real and imaginary parts of the propagation constant β of the fundamental SPP mode [25]. The evanescent decay length quantifies the confinement of the surface wave and is defined as the distance from the interface where the magnitude of the field has dropped to $1/e$ as $\hat{z} = 1/|k_j|$, where j is the metal or dielectric medium of the vector wave. Hence, the larger the decay length, the weaker the confinement will be. Similarly, L_{spp} is the distance at which the intensity of the propagating wave has dropped to $1/e$:

$$L_{spp} = (2\text{Im}[\beta])^{-1}. \tag{3.48}$$

In the infrared region, L_{spp} can take values of several micrometers [49]. Both $\hat{z}$ and L_{spp} depend strongly on the frequency, as is evident from the dispersion relation. The EM wave propagating along the 1D plasmonic waveguide is a slow wave, i.e., it has a smaller phase velocity $v_g = \frac{\partial \omega_{spp}}{\partial k_{spp}} = \frac{c \lambda_{spp}}{\lambda_0}$, a shorter wavelength λ_{spp}, a higher momentum, and a higher wave impedance compared to regular waves in the same medium. This effect is the essence of the nanochip design based on plasmonic waveguides.

In this way, a plasmonic waveguide can also confine in two dimensions, as in Fig. 3.5. In Fig. 3.5a, b, the 2D waveguide is formed by a lateral modification of the dielectric internal space of the gap in the plasmonic waveguides, whereas Fig. 3.5c shows an analog of an RF waveguide, which is primarily a $\lambda/2$-limited device. The other plasmonic waveguides in Fig. 3.5 are open structures, thus supporting radiation as well as leaky modes (in contrary to the 1D gap waveguide). In Fig. 3.5d, e, the slot and trench waveguides can also be considered as transmission lines (two not simply connected metal layers), whereas all other configurations are based on modified single metal surfaces.

In addition, plasmonic waveguides with special geometries allow us to confine light at nanoscale, sub-wavelength dimensions away from the plasmon resonance [50], thus opening the door to both strong coupling with QEs and the construction of compact nanophotonic circuits. However, two problems must first be solved to realize these goals. First, the momentum mismatch between SPPs and free-space photons makes it difficult to couple in-to-plane and out-of-plane SPP modes from the far field. Second, losses in metallic nanostructures limit the propagation lengths for confined modes, making it difficult to construct resonant cavities and integrated circuits. In this sense, some advances have been made using new geometries with lower losses and plasmonic waveguides by incorporating gain media [51–56].

García-Blanco et al. [57] presented a theoretical study of compensation for propagation losses in metal-loaded hybrid plasmonic waveguides with Yb^{3+}-doped potassium double-tungstate gain material. The effects of different structural parameters, height, and refractive index of the buffer layer and the width of the gold stripe were studied. Depending on the structure chosen, compensation for propagation loss and even a net gain were expected from using Yb^{3+}-doped potassium double tungstate as a gain material, opening the door to interesting applications of these types of devices.

Dahal et al. [58] demonstrated a strip of optical waveguides based on Er-doped AlGaN/GaN:Er/AlGaN heterostructures, which were characterized in the optical communication wavelength window near 1.54 μm. The propagation loss of this waveguide amplifier at 1.54 μm was 3.5 cm^{-1}, and a relative signal enhancement of ~8 cm^{-1} was observed.

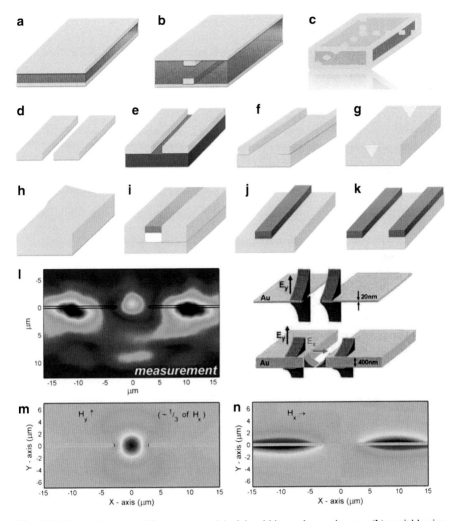

Fig. 3.5 Plasmonic waveguide structures: (**a**) slab within a plasmonic gap, (**b**) variable-size plasmonic gap, (**c**) rectangular channel in metal, (**d**) slot line, (**e**) trench, (**f**) rectangular channel, (**g**) v-groove channel, (**h**) metal wedge, (**i**) partially loaded channel, (**j**) and (**k**) slab and void layers on metal. (**l**) Modal pattern of a 10-μm-wide trench engraved in a 400-nm-*thick gold layer*, embedded in a dielectric host ($n = 1.5$), with the TE (E_x) field excited at $\lambda = 1.55$ μm. The gold layer borders are graphically illustrated. (*l*) Experimental imaging of optical power at the output facet in comparison to an FVHFD-calculated guided mode (second mode). (*m*) E_x/H_y EM field component, with a peak value 3.5 times higher than the (*n*) E_y/H_x EM field components. *Inset*: The symmetrical SPP waveguide structures studied: (*upper*) slot and (*lower*) trench (Pictures (**l**), (**m**), and (**n**) are extracted from ref. [66]). Reprinted under United States Copyright law (Optical Society American)

Despite the importance of this subject, we will not include it in this section of the book. However, there are extensive literature and excellent reviews about this subject; see, e.g., refs. [59–62, 82]. It is worth noting that not all of the plasmonic waveguides shown in Fig. 3.5 are equally capable of guiding sub-wavelength plasmonic signals. For example, in the optical communications region (1,300–1,570 nm), wedge plasmon waveguides have been shown to be superior to groove waveguide structures, because of their strong nanoscale localization of guided plasmonic signals, their relatively low dissipation, and their large L [63]. This suggests that wedge plasmonic waveguides are preferred over v-groove waveguides for nanoscale interconnects in the near-infrared. At optical frequencies, however, is the contrary case, in which v-groove plasmonic waveguides provide stronger field localization and larger L [64] than metal wedges [65].

We must note that not all SPP modes guided by these structures can be used to achieve a nano-localization of the guided signals. For example, metal films [67] or strips [68, 69] can guide either long-range or short-range SPPs, but if the thickness of the film or strip decreases, then we obtain a poorer localization of the long-range mode. On the other hand, not all of the plasmonic waveguides are equally capable of guiding signals. For example, chains of NPs exhibit very strong dissipation [70], making them difficult to use for efficient interconnects. Nanoholes in a metallic surface also exhibit high dissipative losses [71]. The advantages, efficiencies, and optimization of different types of plasmonic and conventional waveguides, as well as interconnects and optical signal processing for nanophotonics have been discussed [72].

3.5 Effects of Nanostructure Arrays on Quantum Emitters

The field of plasmonics has gained great interest in the last two decades, since it has been established that the excitation of plasmons on metallic nanostructures can be usefully exploited to handle several issues in many applications, such as Raman spectroscopy, sensing, data storage, light emission and amplification, as well as absorption enhancement for energy harvesting purposes, and nanodevices for photonics [25, 40, 73]. Today, these investigations provide us with an increasing variety of metallic nanostructures, which can be either resonant or nonresonant, both with extraordinary capabilities and the ability to concentrate light into nano-scale volumes. In resonant nanostructures, time-varying electric fields exert a force on the electron gas inside the metal and drive them into an SPP. At specific optical frequencies, this oscillation is resonant, producing LSPR. Hence, it is produces a very strong charge displacement and an associated field concentration (light). Furthermore, for nanostructures with one or more dimensions approaching λ_{spp}, the optical Φ can vary across the structure and thus it is necessary to consider retardation effects.

The concepts behind "light concentrator" resonant nanostructures are based on scaled radiofrequency antenna designs, such as the dipole antenna. An optical antenna is the equivalent of a classical antenna, i.e., it optimizes the energy transfer between a localized source or receiver and a free-radiation field. However, the term "optical antenna" has clearly been extended beyond its common definition in radio-wave technology. In this case, an optical antenna is not just a resonator or a strong scatterer. Instead, it functions as a transducer between free radiation and localized energy. Therefore, its performance is defined for the degree of localization and the magnitude of transduced energy (EM waves for electric currents or vice versa) [74]. Such scaling is nontrivial since metals display finite conductivity and support SPPs at optical frequencies.

On the other hand, nonresonant nanostructures can also be used to further enhance light concentration. For example, strong nanoscale light localization can be achieved in retardation-based resonators by introducing a small gap in the metallic nanostructure called a feedgap [75]. Here, one takes advantage of the fact that SPPs have an appreciable longitudinal electric field component (normal to the gap) that jumps across the gap (due to the buildup of charges of opposite sign on either side of the gap), given by the ratio $\varepsilon_m/\varepsilon_d$ [76]. Therefore, the local field intensities can be enhanced through "lightning rod" effects [77]. Plasmonic tapers, such as metal cones or wedges, can also provide broadband, nonresonant enhancements. These structures support SPP waves that show an increase in k_{spp} and a decrease in v_g as they propagate toward their tip [68].

Furthermore, in plasmonic nanostructures we can have linear and nonlinear effects when SPPs propagate into metal grating, nanoholes, or slits. That is, under certain circumstances (specific geometrical parameters, wavelengths, and angles of incidence), an enhancement of transmission occurs. Metal gratings composed of sub-wavelength slits permit the formation of a plasmonic bandgap, i.e., a region where transmission is forbidden [30, 78] in the vicinity of the wavelengths where EOT occurs.

The science of photonics covers the generation, emission, transmission, modulation, signal processing, switching, coupling, amplification, and detection/sensing of light. Thus far, we can see a great similarity between photonics and plasmonics, but the greatest difference is the scale. Photonics appeared in the late 1960s to describe a research field whose goal was to use light for purposes that traditionally fell within the typical domain of electronics, such as telecommunications, information processing, etc. Today, it covers all technological applications of light and covers almost the entire EM spectrum, from UV to visible, to the near-, mid-, and far-infrared [79]. Nevertheless, most applications are in the range of visible and near-infrared light.

There are a number of physical systems being investigated for the development of future technologies [80, 81], however, those involving quantum states of light show the most promise. Light is a logical choice for quantum communication, quantum computation, and optical communications, and is a leading approach to quantum information processing. Quantum communication, i.e., a reversible light–matter interface, can be achieved through the direct transfer of quantum states of a QE into matter and back, or through the generation of light–matter interaction in a QE followed by transportation of quantum information from an externally provided photon into matter, and vice versa [54, 82–84]. As mentioned above, wherever nanoscale control over light is desired, metallic nanostructures are likely to play an important part in the development of new technologies for photonics.

At present, we are witnessing a dramatic growth in both the number and scope of plasmonic applications. Standing out among these is the field of nanophotonics, which, beyond the diffraction limit, allows us the propagation, localization, and guidance of strongly localized SPP modes using nanostructures. In this scenario, however, there are still several unresolved issues impeding the tremendous potential of SPP-based nanophotonic circuits, such as efficient coupling to nanoscale waveguides and the implementation of functional devices within the limitations imposed by unavoidable SPP propagation losses. Therefore, the possibility of engineering the optical properties of SPPs and LSPRs in nanophotonics will depend on the efficient photon–plasmon and plasmon–photon conversion, and overcoming the problems mentioned above. An effective solution to the SPP propagation losses (namely strong damping due to the ohmic behavior) is achieved by a gain-assisted medium [55, 85, 86], which can improve the coupling and propagation of these SPPs or LSPR. Here, the QE present in the gain medium can couple with the SPP or with LSPR, i.e., a plasmon–photon interaction, and this can be divided in two regimes, the weak-coupling and strong-coupling regimes.

As discussed in Sect. 1.5, the spontaneous emissions of a QE depend on the crystal field, which can be modified by the presence of metallic NPs or by the vicinity of a nanostructure array. In both cases, this "new" EM environment can also be called cavity quantum electrodynamics (CQED). In this framework, CQED involves the interaction of a QE with a tailored EM field and a high quality factor (Q) and small volumes (V_{eff}), which are related by the Purcell factor [87]:

$$F_{\text{p}} = \frac{\gamma_{\text{em\_cavity}}}{\gamma_{\text{em}_{\text{free-space}}}} \propto \frac{3}{4\pi^2}\left(\frac{Q\lambda^3}{V_{\text{eff}}}\right), \qquad (3.49)$$

where V_{eff} is the mode volume and is given by [88]:

$$V_{\text{eff}} = \left(\frac{\lambda_0}{n_{\text{d}}}\right)^3 \frac{\int \varepsilon|E|^2 dV}{\max\left(\varepsilon|E|^2\right)}. \qquad (3.50)$$

The strength of the interaction between the QE and the surrounding field is defined by a coupling frequency:

$$g = p_{\text{qe}}\sqrt{\frac{\omega}{\varepsilon_0 \hbar V_{\text{eff}}}}. \qquad (3.51)$$

Both interactions (weak coupling and strong coupling) are determined by the CQED, and are dependent on the comparison of g and the damping rates of both the QE and the cavity (γ, k). Therefore, a weak-coupling regime is given by $g \ll (\gamma, k)$, and a strong-coupling regime is given by $g \gg (\gamma, k)$ [89].

The weak coupling is associated with the Purcell enhancement of spontaneous emission (Eq. (3.49)), where the QE couples with the confined SPP or LSPR modes. This weak coupling can have two processes: (1) favorable, where the intense plasmonic field increases the excitation rate of the QE, and (2) detrimental, where the plasmonic field improves the decay rate of the QE into SPPs or LSPR modes via the Purcell effect, i.e., emission quenching. This occurs close to the metallic surface, therefore, an optimal distance for the plasmon–photon coupling appears.

On the other hand, in strong coupling, a reversible exchange of energy, known as Rabi oscillation, occurs between the QE and the cavity field [90]. Rabi oscillations manifest themselves in an energy splitting of the light–matter energy levels. Therefore, both subsystems (emission from the QE and SPP or LSPR-termed quantum-plasmonic), can no longer be treated separately, hence, the interaction of light and matter must be dealt with using a perturbative approach [91, 92]. There are also theoretical studies based on a fully quantum mechanical framework that display the strong-coupling regime [93–95]. Akimov et al. [54] reported the enhanced emission of a single quantum dot into an SPP mode in a silver nanowire (Fig. 3.6a). Additionally, they observed that light scattered from the end of the nanowire was anti-bunched (Fig. 3.6b). This confirming that the SPP mode could collect and radiate single photons from quantum dots, namely acting as a mediator for entanglement between quantum states.

Jin et al. [84] demonstrated quantum entanglement between two quantum dots (QD) in a plasmonic waveguide, shown in Fig. 3.6c, with a near-zero mode index, as functions of concurrence on interdot distance, the frequency detuning, and coupling strength ratio. They observed a high concurrence rate for a wide range of interdot distances owing to the near-zero mode index, which largely relaxes the strict requirements on interdot distance in conventional dielectric waveguides or metal nanowires.

An important detail that has not yet been mentioned is the large size mismatch between a plasmonic nanostructure and a single QE, which ensures that their light–matter interaction will be intrinsically weak. This is a serious problem, since coherent coupling between SPPs/LSPR and QEs is critical for developing future quantum technology [96]. There are several approaches that attempt to resolve this problem, e.g., increasing the Q-factor (Eq. (3.49)), so that the strong-coupling regime can be entered more easily. This can be achieved by reducing the damping of the plasmonic nanostructure by fabricating it from other materials, such as oxide or nitride thin films [97]. Another solution could be the use of a gain medium, with a higher concentration of QEs in transparent substrates [98], i.e., the interaction of multiple QEs mediated through a strong interaction with a plasmonic mode [99].

In Sect. 3.3 of this chapter, we discussed some of the properties of a plasmonic array and saw that it can function as a light scattering concentrator, where each nanohole acts like an antenna or nanocavity, and the spacing and dimensions between them are much smaller than the incident wavelength. These characteristics go beyond the well-established frequency-selective technologies, and are

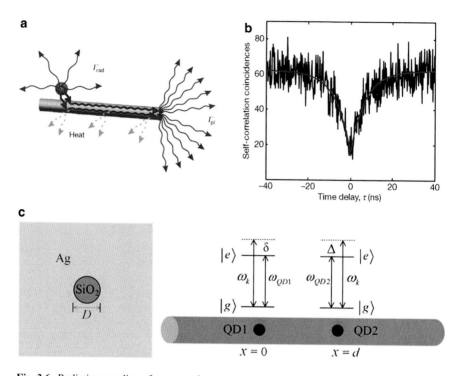

Fig. 3.6 Radiative coupling of quantum dots to conducting nanowires. (**a**) A coupled quantum dot can either spontaneously emit into free space or into the guided surface plasmons of the nanowire with the respective rates γ_{rad} and γ_{pl}. (**b**) Second-order cross-correlation function between fluorescence of the quantum dot and scattering from the nanowire. The *black* and *red traces* indicate experimental data and best fits, respectively. This figure extracted from ref. [54]. (**c**) Quantum entanglement in plasmonic waveguides: (*left*) schematic of the cross section of a SiO$_2$ waveguide with a thick silver cladding, where D is the diameter of the SiO$_2$ core, and a 3-μm-long waveguide is used; (*right*) a pair of two-level QDs separated by distance d interacting with the waveguide mode, where δ and Δ are the frequency detunings between the incident waveguide mode and the QD1 and QD2 transitions, respectively. This figure extracted from ref. [84]. Reprinted under United States Copyright law (Optical Society American)

expanding to new spectral regions, such as the visible and NIR regions [19, 100]. In this manner, coupling between the plasmonic cavity and the QE can enable a host of potential applications related to the development of nanophotonics. One of the advantages of plasmonic cavities is their compatibility with a wide variety of QEs and their broadband cavity spectra, which enable a broad-spectrum enhancement of the QEs [101, 102].

Kim et al. [103] presented a quantum logic gate between a quantum dot (a QE) and a photon. This is an important enabler for robust and scalable quantum networks and for the generation of strong photon–photon interactions (between an optical photon and a solid-state quantum bit—qubit). The qubit is composed of a quantum dot strongly coupled to a metallic nanocavity, which acts as a coherently controllable qubit system that conditionally flips the polarization of a photon on

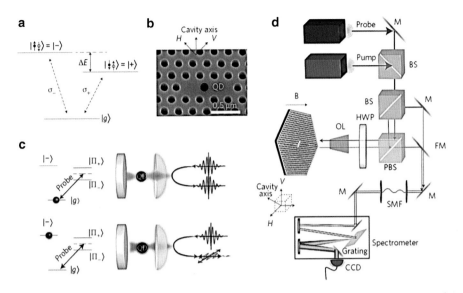

Fig. 3.7 Implementation of a QD-photon controlled-NOT operation. (**a**) Energy-level structure of a neutral QD under a magnetic field. (**b**) SEM image of the fabricated device and the cavity axis relative to photon polarization. (**c**) Illustration of the controlled-NOT operation. The polarization of an incident photon is preserved when the quantum dot is in the state $|g>$ (*top*), and is rotated when the quantum dot is in the state $|->$ (*bottom*). The horizontal dashed lines indicate the degenerate energy levels of the $|+>$ quantum dot state and the cavity photon state, which are split into two polariton states, $|\prod_+>$ and $|\prod_->$, in the strong coupling regime. (**d**) Measurement set-up. Pump and probe polarizations are selected and measured using a polarizing beam splitter (PBS) and a half-wave plate (HWP). A flip mirror (FM) is used to direct the probe signal from either the transmitted or reflected part of the PBS to a single-mode fiber (SMF) and then to a grating spectrometer. Here, *OL* objective lens, *BS* beam splitter, *M* mirror, *QD* quantum dot. This figure extracted from ref. [103]. Reprinted with permission of copyright Nature publishing group

picosecond timescales, thus implementing a controlled-NOT gate, as illustrated in Fig. 3.7. When combined with local quantum dot tuning methods through a plasmonic cavity, these devices provide a potential route toward quantum information processing on a chip.

Lo et al. [104] showed that single-crystalline Er-doped ZnO nanorod arrays on Ag island films with appropriate annealing are a promising device for enhancing 1,540-nm emissions for optical communication. The enhanced 1,540-nm emission of an Er-doped ZnO plasmonic array was attributed to the enhancement of the deep-level emission of the ZnO host. In an effort to enhance deep-level emission to pump Er^{3+} emissions at 1,540 nm in this system, surface plasmon coupling and increased deep-level states were produced via Ag island films and high-temperature annealing.

On-demand single photon sources operating at telecommunication wavelengths are crucial building blocks for fiber-based quantum information networks. In this sense, there is no doubt that Er^{3+}-doped materials have potential for engineering on-chip light sources and laser structures operating at 1.54 μm [100, 105, 106]. Using plasmon–photon interaction, they increase the optical properties in optical telecommunications.

3.5.1 Transmission Enhancement

As demonstrated throughout this book, SPPs are well known for their ability to concentrate light in sub-wavelength volumes, for their EOT, and for guiding light along the surface of a metal. It has also been found that very strong directional emission (beaming) is possible through single apertures if corrugation is placed on the exit side, as discussed in Sect. 3.2. Now, we concentrate on explaining the fundamental physics behind the phenomenon of EOT in a gain medium with an interface (Chap. 1, Sect. 1.3), in a 2D hole array, and in single aperture, and also the physics behind the beaming effect observed in single apertures surrounded by periodic corrugations. Using SPPs or LSPR to confine light gives us the ability to significantly alter the LDOS. In consequence, the dynamics of light–matter interaction can be significantly modified and improved.

Metallic nanostructures sustaining surface plasmons can change the excitation and emission properties of locally excited QEs. Of course, depending on the size and geometry of the nanostructure (a nanocavity or nanoantenna), quantum effects can be significant in the description of the electrodynamics, where the quantum dynamics of emissions from the QEs are also greatly modified by coupling to a cavity, with increased decay rate and emission intensity. Thus, plasmonic nanostructures act as physically realizable classical oscillator systems at the nanoscale, with potential applications and benefits to designing models for a variety of fascinating physical processes.

As mentioned above, the most common valence state of REIs in solids is the trivalent state; see Chap. 2. In this state, the 4f electrons are weakly perturbed by the charges of the surrounding atoms, and thus the wavelengths of the emission/ absorption transitions are relatively insensitive to the host material. Nevertheless, the line shapes of the emission spectra and the emission intensity depend on its surroundings. Let us suppose the presence of an REI near a plasmonic cavity. In the interaction process, this cavity can profoundly change the number of EM modes and decay pathways available to an emitter, increasing or decreasing the radiative rate of the QE. Therefore, we can manipulate the LDOS—namely, the EM environment (the crystal field) of an REI. These changes in the LDOS have been used to increase the efficiency of light-emitting diodes, the intensity of atomic and molecular fluorescence, and even the likelihood of forbidden transitions, such as optical-frequency magnetic transitions [107, 108].

In this manner, the rate of spontaneous emission of a QE in the weak-coupling regime is given by Fermi's rule:

$$\gamma = 2\pi g^2 \int \rho_e(\omega)\rho_c(\omega)d\omega, \tag{3.52}$$

where g is the coupling between the QE and the EM field, $\rho_e(\omega)$ is the LDOS of the QE at ω, and $\rho_c(\omega)$ is the LDOS of the EM environment. Equation (3.52) is similar to Eq. (1.34), but here we consider the changes in the LDOS due to the coupling between the QE and the plasmonic cavity. The Rabi frequency is given by Eq. (3.51). Typically, the dielectric constant in a dispersive medium is substituted by $\varepsilon(\omega) = n_d^2$, e.g., for a particular transparent medium we can use the Sellmeier equation, which is an empirical relationship between n_d and ω. Therefore, the spontaneous emission rate of a QE is proportional to the LDOS, resulting in increased emission owing to the changes in the LDOS of the QE, and also an increase in the $\gamma_{em\_cavity} = \gamma_{QE} + \gamma_{SPP}$. A large LDOS can increase not only the γ (Eq. (3.52)), but also the stimulated emission process within the plasmonic cavity, namely a lasing action without requiring stronger pump power [109].

On the other hand, a plasmonic array can be treated as Bloch modes (Sect. 3.3.1) and can sustain a large LDOS, because its intense optical fields are coupled to oscillating charges in the metal nanostructure [110]. In this way, while this plasmonic array possesses a countless number of mirror symmetry wave-planes, one cavity can possess only one of them. Therefore, we can enhance the emission of one nanocavity by this plasmonic array. This means that the study of the LDOS in nanostructures on a gain medium is essential for understanding and developing novel photonic nanodevices.

A quantum treatment of SPPs and LSPR here can be a useful tool for describing the microscopic interaction between light and matter. Such a quantum treatment allows us to model stimulated emission and, therefore, specify gain conditions and laser operations in a quantum–plasmonic interaction. For this, we start from the premise that an SPP (or LSPR) can store high values of energy in small volumes. Then, we use a quantization scheme for EM fields in dispersive and absorptive media introduced by Gruner and Welsch [111]. Therefore, the EM energy associated with surfaces waves is given by [91]:

$$U = \sum_k \varepsilon_0 \omega^2 S \left(A_k A_k^* + A_k^* A_k \right). \tag{3.53}$$

where A_k is the amplitude of the vector potential dependent on position and time and $S = s(r, t)$ is the displacement of a small volume of charge inside a plasma characterized by carriers with charge e, mass m, and density N [55]. We see that the expression of energy is for each mode k, and has the structure of the energy of a harmonic oscillator, and that the quantized Hamiltonian is [91]:

$$\hat{H} = \sum_k \frac{\hbar \omega_k}{2} \left(\hat{a}_k \hat{a}_k^\dagger + \hat{a}_k^\dagger \hat{a}_k \right), \tag{3.54}$$

with the relations:

$$A_k \rightarrow \sqrt{\frac{\hbar}{2\varepsilon_0 \omega_k S}} \hat{a}_k \tag{3.55}$$

and

$$A_k^* \rightarrow \sqrt{\frac{h}{2\varepsilon_0 \omega_k S}} \hat{a}_k^\dagger. \tag{3.56}$$

The SPP (or LSPR) is thus quantized by association with a quantum-mechanical harmonic oscillator to each mode k. This also introduced $\hat{a}_k^\dagger$ and $\hat{a}_k$, i.e., the creation and annihilation operators for mode k. Both operators (as in harmonic-oscillator theory) allow us to create or annihilate, respectively, a quantum of energy $h\omega_k$ according to the operator rules [112]:

$$\hat{a}_k|n_k\rangle = \sqrt{n_k}|n_k - 1\rangle \tag{3.57}$$

and

$$\hat{a}_k^\dagger|n_k\rangle = \sqrt{n_k + 1}|n_k + 1\rangle. \tag{3.58}$$

Different SPP (or LSPR) modes are independent, i.e., their associated operators commute: $\left[\hat{a}_k, \hat{a}_{k'}^\dagger\right] = \delta_{k,k'}$.

Assuming that the REIs may occupy different sites in the host matrix (a dielectric), a direct coupling between the excited states of the REIs and the plasmonic nanocavity modifies the Stark level energies. The nanocavity can only contribute to the local field when it is excited, so the oscillator strength (P_{strength}) of a spectral line, corresponding to a transition from the REI's ground level, i, to a component, f, of the excited level, is given by:

$$P_{\text{strength}} = \chi \left[\frac{8\pi^2 m\nu}{3h(2J+1)}\right] \left\{\sum |\langle i|D_{\text{q\_nearest-neighbor}}|f\rangle|^2 + \sum |\langle i|D_{\text{q\_nanocavity}}|f\rangle|^2\right\}, \tag{3.59}$$

where h is the Planck constant, ν is the frequency of the line, and $q = 2, 4$, and 6. The second term in the parentheses in this equation has been modified from the theory of Judd [113], and represents the ED transition due to the LSPR modes of the nanocavity. The factor χ is an adjustable factor that depends on the refractive index of the medium in which the REI are embedded, and J is the total angular momentum. In order to obtain nonvanishing matrix elements of the components $D_{\text{q\_nanocavity}}$, it is necessary to form an admixture of $\langle i|$ and $|f\rangle$ with other states of opposite parity. This means that the role of the nanocavity is not related to the variation of the characteristics of the wave functions (even or odd), but instead is related to the transition intensities and probabilities within the 4f shell, which can be evaluated from Eq. (3.51).

In order to experimentally demonstrate improved EOT in both single and array nanocavities with REIs, we can design a bull's-eye pattern with a corrugated surface [28], where an emitting dipole (an Er^{3+} ion QE) positioned in proximity to a metallic surface decays nonradiatively, on a fast time scale, by interaction with the LSPR modes of the system. Afterward, the LSPR modes can decay (out-couple) as radiative modes or by nonradiative decay, depending on the ratio of the albedo of the plasmonic nanostructure to the measured wavelength. Using this design, metallic nanostructures with tunable surface plasmon resonances, and the coupling of Er^{3+} ions to such structures, might enable the discovery of interesting optical and electronic properties. We also find enhanced EOT with REIs in different plasmonic arrays [19, 28, 100–102] in the visible and near-infrared regions, where a high increase in efficiency was achieved thanks to the enhanced and localized transmittance in each nanocavity. These experimental works and related quantum-plasmonic models provided a stimulating backdrop for researchers as they began to explore plasmonic systems using quantum optical techniques.

3.5.2   Focusing of Surface Plasmon Polaritons

Lens provide us with a convenient way to manipulate light, by transmitting, refracting, converging, and diverging, and also improving a range of linear and nonlinear phenomena, and they are typically made of glass or transparent plastic. However, these dielectric lenses cannot focus light to spots less than about half a wavelength of light ($\lambda/2$), and dielectric resonators have EM mode volumes, V_{eff}, limited to $V_{eff} \approx (\lambda/2)^3$. However, the nature of plasmonic nanostructures allows us to move beyond these limits (the fundamental laws of diffraction), since these nanostructures can concentrate and resonate light at the nanoscale (Eq. (3.50)), where peculiar SPP/LSPR properties are exhibited.

Only a short time ago, focusing light to the nanoscale was a technological challenge, but today's developments in SPPs allow for the future research in nanophotonics. In this case, nanofocusing of SPPs can occur when optical energy is compressed and concentrated into the nanoscale. Nanofocusing of SPPs can be achieved using metallic nanocavities with various configurations, such as nanoholes, nanoslits, nanorods, nanotips, wedges, grooves, nanorings, and NP chains. These are the structures commonly used in SNOM. In this manner, nanofocusing is one of the major tools for efficient coupling light and light-carried information into nano-waveguides, interconnectors, and optical nanodevices [114]. Such focusing will also aid in improving the resolution of microscopes [115, 116]. All the nanostructures mentioned above—except for wedges—provide equally strong field localizations in regions as small as a few nanometers, limited only by spatial dispersion and the atomic structure of matter [117].

Plasmon nanofocusing can occur in an adiabatic regime or a nonadiabatic regime. In the adiabatic regime, i.e., in a metal wedge, plasmon nanofocusing may only occur if the wedge angle is smaller than a critical angle determined by the dielectric permittivities of the metal wedge and the surrounding dielectric medium [118, 119]. If the taper angle is increased further, then significant reflections of the plasmons off the taper occur. Therefore, there exists an optimal taper angle for achieving maximum enhancement near the tip of the focusing structure, and also an optimal length of the taper [120–122]. In the nonadiabatic regime, the reflective energy losses (dissipative losses) in the plasmon as it propagates toward the tip may be noticeable and should be taken into account, resulting in a reduction of the local field enhancement near the tip [121, 122].

A conical metal tip is also capable of focusing light to a nanoscale spot, allowing the accumulation of energy and giant local fields at the tip. It gives rise to applications in sensing, microscopy, and the efficient coupling of light to nanodevices [123, 124].

On the other hand, Jennifer M. Steele et al. reported the generation and focusing of SPP modes from normally incident light on a planar circular grating milled into a silver film [31]. They explained their results using a simple coherent interference model for SPP generation on a circular grating by the incident light. Experimental and theoretical results predict and highlight the requirement for phase matching of SPPs in the grating to achieve maximum enhancement of the SPP wave at the focal point, i.e., the center of rings. SNOM measurements show that the plasmonic lens achieves a more than tenfold increase in intensity over that of a single ring when the ring-groove design is tuned to the SPP wavelength. For instance, Kim and Cheng numerically presented the increased optical field of gap SPPs based on a circular slit at a metal–dielectric interface [125]. With the combination of gap SPPs and propagating SPP matching by the circular plasmonic lens to achieve a strong field confinement, such a nanostructure can be expected to be applied to high-sensitivity molecule detection by surface-enhanced Raman scattering.

Owing to the strong confinement by the metallic nanostructures, we can profoundly alter the light emission properties of nearby QEs [28, 126–128], i.e., we can manipulate radiative decay by: (1) increasing optical excitation rates, see Eq. (3.52); (2) modifying radiative and nonradiative decay rates, see Eq. (3.54); and (3) altering emission directionality. Figure 3.8a, c show SEM images of bull's-eye nanostructures (also called a plasmonic planar lens) milled on a 240-nm-thick gold thin film, which was deposited on an Er^{3+}-doped tellurite glass [28]. Also shown are dark-field microscopic images (Olympus BX53 with a darkfield dry condenser 10× NA 0.8, objective Olympus UPlanSApo 60×/1.2 W NA 0.9—NPLan) obtained from the same samples shown in Fig. 3.8a, c. In Fig. 3.8b, an optical image of Fig. 3.8a, we see a high field concentration (nanofocusing) in the bull's-eye center. Ring-grooves ($h = 80$ nm) show the green luminescence from the Er^{3+} ions. The metallic rings show a red color; this is due to multiple interactions between the SPP and the gold surface within the cavity formed between two rings, as in Fig. 3.8d, and we also see the nanofocusing effect in the bull's-eye center. Both optical images outside the bull's-eye have a black background.

Therefore, this focusing allows a high confinement in the bull's-eye center that excites the Er^{3+} ions, i.e., EOT in the nanohole, incremented through the ring-grooves. The experimental setup for measuring the intensity of the luminescence consists of a 488.0-nm wavelength light beam from an Ar ion laser, with a power of approximately 40 μW and a spot size of 1 μm, aligned to the optical axis of a

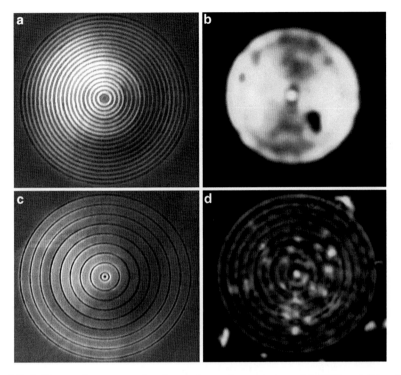

Fig. 3.8 Bull's-eye ring-grooves forming a planar plasmonic lens: (**a**) SEM and (**b**) optical images for a gap of 300 nm with 15 ring-grooves, and (**c**) SEM and (**d**) optical images for a gap of 600 nm with 9 ring-grooves

microscope. The spectra of the bull's-eye as functions of the gap (separation between ring-grooves) are shown in Fig. 3.9. The spacing between the rings determines the phase mismatch between the waves generated from the inner and outer rings. Therefore, the nanohole transmitted the light into the substrate and excited the Er^{3+} ions with an improved intensity, generating an SPP in the gold–substrate interface. The excitation of the Er^{3+} ions depends on their localization to the gold film and of their proximity to the nanohole.

The measured microluminescence clearly confirms that planar plasmonic lenses can focus the excitation beam with great efficiency. We can see that it changes the γ (Eq. (3.52)) and therefore changes the oscillator intensity (Eq. (3.59)). In the inset of Fig. 3.9 is shown the behavior of the intensity of the peaks from different planar lens normalized with the intensity of the nanohole (a square of 500×500 nm^2), for the peaks at 524 nm ($^2H_{11/2} \rightarrow {}^4I_{15/2}$ electronic transition), 547 nm ($^4S_{3/2} \rightarrow {}^4I_{15/2}$ electronic transition), and 659 nm ($^4F_{9/2} \rightarrow {}^4I_{15/2}$ electronic transition). Those curves have a good agreement with Eq. (3.4).

Other nanocavities employed in the field of nanofocusing are concave and convex plasmonic lenses, shown in Fig. 3.10a, b, respectively. Both can enhance the oscillator strength of the Er ion [126], however, we see that they do not have the

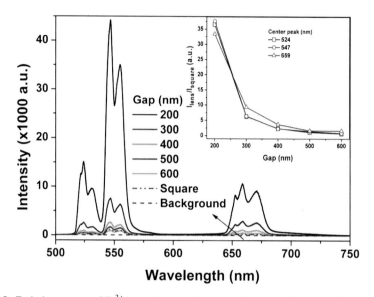

Fig. 3.9 Emission spectra of Er^{3+} ions via the bull's-eye as functions of the gap. We can see that the background line is practically neglected. It also shows the emission spectra of the Er^{3+} ions through a square of 500×500 nm^2. The inset shows the behavior of this increment normalized for the different Er^{3+} ion transitions. These results were extracted from ref. [28]

same coupling efficiency, as shown in the inset of Fig. 3.10c. Figure 3.10c shows the Er^{3+} luminescence through the concave/convex plasmonic lenses, their dependence on the type of lens, and the separation distance between the ring-grooves, in order to obtain an efficient excitation of the substrate. These behaviors are particularly important for understanding the fundamental properties of these structures, as well as their functionalities, in excitation and emission of Er^{3+} ions for applications in telecommunications.

Plasmonic nanocavities exhibit only reasonable values of Q (from approximately 10 up to 100) in comparison with typical CQED. However, the v_{eff} value can be much smaller than a typical dielectric cavity ($v_{\text{eff}} \approx (\lambda/2)^3$), and the resulting Purcell factor (Eq. (3.49)) can be large enough to produce significant lifetime reductions. Mostly large enhancements are obtained when the QE emission frequency matches the LSPR frequency of the nanocavity, as shown by Eqs. (3.55), (3.56), and (3.59), and by Figs. 3.9c and 3.10c. Moreover, once excited, LSPR can either decay nonradiatively because of internal damping, or re-radiate into free space, which is a process detrimental to QE emission. Depending on a variety of parameters, the focusing (QE–nanocavity coupling) can exhibit either increased or decreased (quenching) luminescence, see inset of Fig. 3.10c [130].

From this experimental evidence, we can assume that the QE structure remains unchanged near the plasmonic nanocavity. With nanostructures with sufficiently high v_m, it becomes possible to reach a regime with coherent interactions between an SPP/LSPR mode and a QE. Then, the total decay rate is $\gamma_{\text{em\_cavity}} = \gamma_{\text{QE}} + \gamma_{\text{SPP}}$,

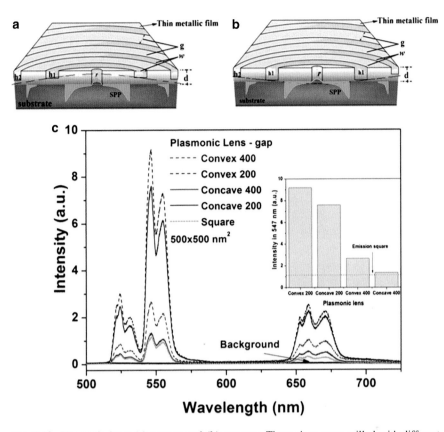

Fig. 3.10 Plasmonic lens: (**a**) convex and (**b**) concave. These rings were milled with different gaps, forming grooves. (**c**) Emission spectra of Er^{3+} ions through a nanohole 200 nm in diameter, for the different plasmonic lens. We can see that the background line is practically neglected. Also shown are the emission spectra of a nanosquare of 500×500 nm^2 (*short dashed line*). The *inset* figure shows the intensity of the central peak for the $^2H_{11/2} \rightarrow {}^4I_{13/2}$ transition of the Er^{3+} compared with the emission intensity for a square of 500×500 nm^2 (*dashed line*). These results were extracted from ref. [129]

increased as a result of the large LDOS near the metal surface. This can also affect the radiative decay from the QE by introducing new EM decay pathways, i.e.:

$$\tau_{\text{uncoupled}} = \left[\frac{1}{\tau_r} + \frac{1}{\tau_{nr}} \right]^{-1}$$

and

$$\tau_{coupled} = \left[\frac{F_p}{\tau_r} + \frac{1}{\tau_r} + \frac{1}{\tau_{nr}} \right]^{-1}. \tag{3.60}$$

The rate $\frac{F_p}{\tau_r}$ is the Purcell enhanced spontaneous emission rate, and τ_r and τ_{nr} are the radiative and nonradiative decay time of the QE, respectively.

From these effects, we have the capacity to create a new type of surface-emitting laser that combines the advantages of photonic lasers (e.g., directional beam emission) with those of plasmonic nanolasers (e.g., nanoscale optical confinement and ultrafast response). Based on the promising future of these nanocavities in a gain medium, we anticipate an outstanding role for REIs in the new generation of optoelectronic nanodevices.

Another application of this SPP nanofocusing technique can be found in surface-enhanced Raman scattering (SERS), a very useful technique in the detailed spectroscopic detection of the vibrational levels of a given analyte. Recently, SERS has been transformed into a potent analytic technique [131, 132] owing to advances in nanofabrication and an increased understanding of plasmonic properties. The sensitivity of the SERS depends upon the fourth power of the local field, and such an extremely high field is possible between NPs that are very near to one another, because the local field due to plasmon excitations is greatest in regions with high local curvature, the so-called hot spots. In this framework, SERS enhancement factors of greater than 10^6 (Eq. (2.47)) have been reported by numerous researchers. This is sufficient enhancement for single or few-molecule detection, and therefore SERS is an extremely useful technique for physical, chemical, and biomolecular detection [133, 134]. Because of the limited content of this book, we recommend that readers find more information about SERS in the works of Halas et al. [131], Morton et al. [132], and McNay [134].

3.6 Polarization Control

Typically, polarizers are made from birefringent materials or grid plates, which are used for the conversion of polarization states [135], as half-wave and quarter-wave plates, for instance. Such materials enable a change of polarization direction and the transformation of a linearly polarized light wave to a circularly polarized wave, respectively. These properties arise from the different propagation velocities of the medium, owing to the nature of the refractive index. Metamaterials have also presented an extraordinary capacity to manipulate the polarization state of light [136]. Furthermore, the implementation of nanostructures, combined with REIs into a substrate, also enables an increased capacity to control the polarization by adjusting some parameters in the experiment, like the shape and dimensions of the nanostructure. A planar 2D array of nanocavities, for instance, can exhibit great dichroism.

LSPR arises from the interaction between NPs embedded into a substrate and a metallic nanostructure array that enhances the local electric field on the nanostructure, where there is near-field coupling between closely spaced particles. Nanoclusters of noble metals hosted in glasses [137] or even in semiconductors [138] exhibit strong absorption of visible light. As we have seen in the previous sections, the size and shape of the NPs that constitute such clusters, and furthermore the dispersive properties of the medium, are critical factors that characterize the enhancement of the local electric field in the vicinity of the NPs. Over the last decade, growth techniques for metallic nano-clusters have been widely applied [139] to the development of plasmonic devices, owing to the ease of creating and controlling the geometry/shape of the NPs. Although the use of nano-clusters for such devices is common, it does not offer great plasmon coupling between metal and glass, since the charge density is low in these specific spots in the host medium. Therefore, employing nanostructure arrays [140], which have been milled on metallic thin films, has been demonstrated to be highly promising for the development of new plasmonic devices, providing nanometer-scale resolution for light focusing, great applications in the spectral analysis of molecules, and the sensing of other complex compounds, for instance [141].

Achieving an enhanced electric field from LSPR modes in metallic nanostructure arrays is more effective compared to that generated by the conventional method, using a metallic thin film on a dielectric surface, as was seen in Chap. 1. This feature occurs because in an array configuration it is possible to reproduce effects of well-known devices, like waveguides, lenses, and antennas [142], for instance. When the local electric field is enhanced by these nanostructures, there is consequently an increase in the excitation and emission rates of the QE, because the radiation pattern from the localized plasmon provides efficient coupling between the emission (photon generation) of the radiation array and the emission from the QE. Since the geometry of the element that constitutes the array modifies the features of the radiation pattern [143], it is important to choose the best format for the nanostructure. Nanolithography techniques, such as a focused ion beam [102, 144], are excellent experimental methods for the development of suitable array sets.

It is possible to regard the periodic metallic nanostructure array as a nanoantenna set. The mode of each nanoantenna determines how the polarization of the emitted radiation from the structure is controlled [145]. The far-field mode defines the polarization of the source excitation, depending on the geometry or shape of the element that constitutes the array. Additionally, the near field converts the polarization of the external field to a new and different field related to the evanescent wave, like a localized plasmon.

In order to achieve a higher intensity of the emitted radiation (far field or near field) from the RE into the substrate, the implementation of a nanostructured array is necessary. Since the radiation pattern of a QE is generally dipolar, coupling between these systems to a set of nanoantennas is efficient because the radiation pattern of the nanostructure matches with that of the QE. Quantitatively, such a matching is represented by the collection efficiency η_0, which was defined in Sect. 2.4.2 in Chap. 2. Furthermore, the increased radiation pattern displays a directionality D' in its emission from the nanoantenna array, which defines the shape of the pattern:

$$D' = 4\pi \frac{P(\theta, \varphi)}{\displaystyle\int P(\theta, \varphi)d\Omega}, \tag{3.61}$$

where θ and φ are the polar and azimuthal angles, respectively, and $P(\theta, \varphi)$ is the power of the set of nanoantennas, which is evaluated from the Poynting vector integrated over the area covered by the array. The Poynting vector is generally obtained from the vector product of the electric and magnetic fields resulting from the coupling of the nanoantennas and the QE, which can be calculated from Maxwell's equations. The denominator considers the power distribution over all solid angles $d\Omega$ that the nanoantennas can reach. Since D' is related to the emitted radiation from the nanostructure array, then, in order to describe the coupling between the RE and this array, it is necessary to define a quantity that expresses the increase in the emitted radiation owing to the presence of the gain material in the substrate. The relation below represents such an increase:

$$G = \eta_0 D', \tag{3.62}$$

where G defines the gain in the emission of the radiation from the nanostructures, which is proportional to the collection efficiency η_0. The enhancement of the local electric field owing to the nanoantenna array also provides greater emission rates from electronic level of the QEs. Since the coupling between the REIs and the nanoantennas is coherent, i.e., the features related to their polarization are similar, such a system can function as a polarization selector, because only a specific output of the electromagnetic wave may emerge from the nanostructure and be detected afterward. Therefore, the configuration created by the nanostructure and the QE can also be used as a polarizer in the nanoscale.

An NP can act as a nanoantenna, but it does not change the directivity because the induced dipole moment r_0 points in the same direction as the REI dipole, and because the two dipoles are closely spaced.

In order to improve the control of polarization, we consider an incident light beam on the nanostructure, defined in the 2D case by the following equation:

$$E_i = \left(E_x \hat{x} + E_y \hat{y}\right) e^{i(k_o z - \omega t)}. \tag{3.63}$$

In this example, the z-axis is the direction chosen for the propagation of the beam. In general, for isotropic materials, after reflection the light wave is characterized by the expression below:

$$E_{r'} = r'\left(E_x \hat{x} + E_y \hat{y}\right) e^{-i(k_o z + \omega t)}, \tag{3.64}$$

where $'r$ is the reflection coefficient. For an anisotropic medium, however, there are different reflection coefficients, $'r_x$ and $'r_y$, related to the two incident waves that are polarized along the directions x and y. Thus, we write the reflected wave for the anisotropic medium as:

$$E'_r = \left({'r_x}E_x\hat{x} + {'r_y}E_y\hat{y}\right)e^{-i(k_o z + \omega t)}. \tag{3.65}$$

The polarization state is controlled when we vary the coefficients ${'r_x}$ and ${'r_y}$, or, additionally, when we change the amplitudes E_x and E_y. We will study both cases.

In the example of a linearly incident EM wave, $E_x = E_y$, and we suppose that it is possible to tune the material parameters such that $\frac{'r_x}{'r_y} = -1$. Thus the polarization direction of the reflected wave is $-\hat{x} + \hat{y}$, which is perpendicular to the original wave with a polarization of $\hat{x} + \hat{y}$. The crucial parameter to control the coefficients ${'r_x}$ and ${'r_y}$ is the phase $\Delta\varphi$, which is associated with the coefficients by the following equation:

$$'r_{x(y)} = e^{i\Delta\varphi_{x(y)}}. \tag{3.66}$$

Such a phase depends on the dispersion properties of the materials that constitute the nanostructure (the metallic thin film and/or the glass).

Furthermore, the geometry of the nanostructure array also provides a means of controlling of the amplitudes of the electric field, where it is possible to achieve cases in which $E_x > E_y$, for example. A nanocavity with an elliptical shape, fabricated at the center of the nanostructure, can be designed to control the polarization state of the output light. In this example, the modes from the nanocavity are expressed by the product of even and odd functions from the radial part of the Mathieu functions [146, 147]:

$$E_x = \begin{cases} Je_m(\xi, q) \cdot Ce_m(\eta, q), m \geq 0 (\text{even}) \\ Jo_m(\xi, q) \cdot Se_m(\eta, q), m \geq 1 (\text{odd}) \end{cases}, \tag{3.67}$$

where $q = \frac{f^2 k_{gsp}^2}{4}$ is a gap function of the wave vector k_{gsp} and the focal length, f, of the nanostructure, $Ce_m(\eta, q)$ with $m \geq 0$ and $Se_m(\eta, q)$ with $m \geq 1$ are two independent classes of even (e) and odd (o) Mathieu functions of the first kind [148], and ξ is defined by the geometry of the nanocavity. Therefore, the electric component of the incident and reflected waves can be projected inside the nanocavity in two parallel directions, on the major axis of the nanocavity. The difference between the modes with resonance frequencies e_{11} and o_{11} allows a phase shift between these two components from the electric field. A phase shift of π is obtained if we consider the nanocavity as the Lorentz oscillator; thus, the phase delay can be expressed by the equation:

$$\delta(\omega) = \tan^{-1}\left[\frac{\gamma\omega}{\omega_0^2 - \omega^2}\right]. \tag{3.68}$$

In addition to the features of the nanocavity, in order to control the polarization state, it is also important to analyze the glass that constitutes the substrate where the nanostructure will be fabricated. For instance, in REI-doped glasses, it is reasonable to describe their optical anisotropy resulting from the presence of such ions. The crystalline potential (Eq. (1.40)), owing to the positions of the atoms that form the lattice in the host matrix, is regarded as a perturbation that splits the energy levels from the REI. Since the atoms of the lattice are bounded by electrical forces, the perturbative potential generates an electric field that alters the features of the radiative emissions of the RE, like the Stark effect treated in quantum mechanics theory. We consider the probability of spontaneous emission of an REI starting from an excited state J, which is defined by [149]:

$$A_{total} = \frac{\Sigma_q \Sigma_J A_q (\varphi_i - \varphi_f)}{3},$$ (3.69)

where the index q represents the polarization state of light emitted by the REI. The numerator is written as a function of the correction of perturbation levels:

$$A_q(\varphi_i - \varphi_f) = \sum_{\beta} \left\{ \frac{\langle \varphi_i | V_{crys} | \varphi_\beta \rangle \langle \varphi_\beta | P | \varphi_f \rangle}{(E_i - E_\beta)} + \frac{\langle \varphi_i | P | \varphi_\beta \rangle \langle \varphi_\beta | V_{crys} | \varphi_f \rangle |}{(E_f - E_\beta)} \right\}.$$ (3.70)

Thus, we notice from the expression above that when we control the polarization state of the excitation light P, we can control part of the polarization state of the light emitted by the REI. The remaining features depend on the crystalline potential, associated with the REI embedded in the host matrix.

3.7 Conclusion

As shown in the various subsections of this chapter, plasmonic nanofocusing structures produce a strong enhancement and confinement of a local field, which has the potential for the development of next-generation devices such as sensors and detectors, and nano-imaging techniques. This may lead to new optical manipulation methods for nanoscale optical communication, biophotonics, clean energy, medical testing, and efforts to integrate these fields. In this manner, new plasmonic nanocavity designs that support dramatic enhancements of spontaneous emission may enable nanoscale light sources that are brighter and faster than the laser sources employed today. Furthermore, the dream of a nanolaser can be realized in the form of the SPASER, or surface plasmon amplification by stimulated emission of radiation [109, 150–152], which may lead to new applications in nanoscale lithography.

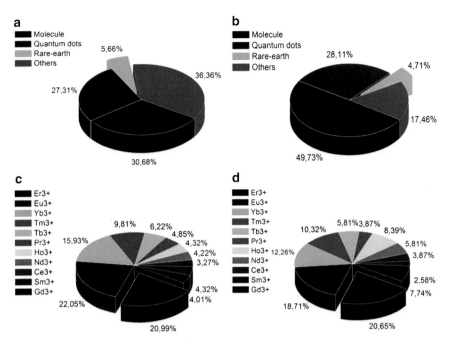

Fig. 3.11 International science and technology related to plasmon modes in gain media: (**a**) global distribution of research on plasmonic NPs (16,759 papers) and the main research areas in this theme and (**b**) global distribution of research on plasmonic nanostructures (3,294 papers) and the main research areas in this theme; (**c**) literature on plasmonic NPs and REIs and (**d**) literature on plasmonic nanostructures and REIs. The REIs displayed here are those most found in the literature. Data sourced from SCOPUS (as of April 2014)

However, we note that research into quantum–plasmonic interactions using REIs is still embryonic, as shown in Fig. 3.11, but should receive a boost because of its various current technological applications. In this sense, we hope we have provided a valuable perspective on the field and that it will help inspire the next trends of plasmonic science and technologies.

As described throughout this book, the field of plasmonics has grown dramatically over the last two decades and new directions are continuously emerging from the field of quantum plasmonics. This chapter has provided only a brief overview of the interaction between QEs and nanocavities, which is an exciting branch in plasmonics with promising technological applications.

References

1. Ebbesen, T.W., Lezec, H.J., Ghaemi, H.F., Thio, T., Wolf, P.A.: Extraordinary optical transmission through sub-wavelength hole arrays. Nature **391**, 667 (1998)
2. Weiner, J.: The electromagnetics of light transmission through subwavelength slits in metallic films. Opt. Express **19**(17), 16139 (2011)

3. Yin, L., Vlasko-Vlasov, V.K., Rydh, A., Pearson, J., Welp, U., Chang, S.-H., Gray, S.K., Schatz, G.C., Brown, D.B., Kimball, C.W.: Surface plasmons at single nanoholes in Au films. Appl. Phys. Lett. **85**, 467 (2004)
4. Rindzevicius, T., Alaverdyan, Y., Sepulveda, B., Pakizeh, T., Kall, M., Hillenbrand, R., Aizpurua, J., de Abajo, F.J.G.: Nanohole plasmons in optically thin gold films. J. Phys. Chem. C **111**, 1207 (2007)
5. Park, T.-H., Mirin, N., Britt Lassiter, J., Nehl, C.L., Halas, N.J., Nordlander, P.: Optical properties of a nanosized hole in a thin metallic film. ACS Nano **2**(1), 25 (2008)
6. Yuan, Z., Gao, S.: Linear-response study of plasmon excitation in metallic thin films: layer-dependent hybridization and dispersion. Phys. Rev. B **73**, 155411 (2006)
7. Bethe, H.A.: Theory of diffraction by small holes. Phys. Rev. **66**(7–8), 163 (1944)
8. Bouwkamp, C.J.: On Bethe's theory of diffraction by small holes. Philips Res. Rep. **5**(5), 321 (1950)
9. Tan, H.S.: On Kirchhoff's theory in non-planar scalar diffraction. Proc. Phys. Soc. **91**, 768 (1967)
10. Gordon, R., Brolo, A.G.: Increased cut-off wavelength for a subwavelength hole in a real metal. Opt. Express **13**(6), 1933 (2005)
11. Shin, H., Catrysse, P.B., Fan, S.: Effect of the plasmonic dispersion relation on the transmission properties of subwavelength cylindrical holes. Phys. Rev. B **72**(8), 085436 (2005)
12. García, N., Bai, M.: Theory of transmission of light by subwavelength cylindrical holes in metallic films. Opt. Express **14**(21), 10028 (2006)
13. García-Vidal, F.J., Moreno, E., Porto, J.A., Martín-Moreno, L.: Transmission of light through a single rectangular hole. Phys. Rev. Lett. **95**(10), 103901 (2005)
14. García-Vidal, F.J., Martín-Moreno, L., Ebbesen, T.W., Kuipers, L.: Light passing through subwavelength apertures. Rev. Mod. Phys. **82**, 729 (2010)
15. de Abajo, F.J.G.: Colloquium: light scattering by particle and hole arrays. Rev. Mod. Phys. **79**, 1267 (2007)
16. Weiner, J.: The physics of light transmission through subwavelength apertures and aperture arrays. Rep. Prog. Phys. **72**(6), 064401 (2009)
17. Ferri, F.A., Rivera, V.A.G., Osório, S.P.A., Silva, O.B., Zanatta, A.R., Borges, B.V., Weiner, J., Marega Jr., E.: Influence of film thickness on the optical transmission through subwavelength single slits in metallic thin films. Appl. Opt. **50**(31), G11 (2011)
18. Bozhevolnyi, S.I., Erland, J., Leosson, K., Skovgaard, P.M.W., Hvam, J.M.: Waveguiding in surface plasmon polariton band gap structures. Phys. Rev. Lett. **86**, 3008 (2001)
19. Rivera, V.A.G., Ferri, F.A., Osório, S.P.A., Nunes, L.A.O., Zanatta, A.R., Marega Jr., E.: Luminescence enhancement of Er^{3+} ions from electric multipole nanostructured arrays. Proc. SPIE **8269**, 82692H–1 (2012)
20. Zayats, A.V., Smolyaninov, I.I.: Near- field photonics: surface plasmons polaritons and localized surface plasmons. J. Opt. Pure Appl. Opt. **5**, S16 (2003)
21. Smolyaninov, I.I., Zayats, A.V., Stanishevsky, A., Davis, C.C.: Optical control of photon tunneling through an array of nanometer-scale cylindrical channels. Phys. Rev. B **66**, 205414 (2002)
22. Osório, S.P.A., Silva, O.B., Ferri, F.A., Rivera, V.A.G., Zanatta, A.R., Marega Jr., E.: Integrated hybrid plasmonic cavity with in-plane photon-plasmon coupling for luminescence enhancement. Proc. SPIE **8269**, 826928–1 (2012)
23. Kretschmann, M., Maradudin, A.A.: Band structures of two-dimensional surface-plasmon polaritonic crystals. Phys. Rev. B **66**, 245408 (2002)
24. Hooper, I.R., Sambles, J.R.: Dispersion of surface plasmon polaritons on short-pitch metal gratings. Phys. Rev. B **65**, 165432 (2002)
25. Raether, H.: Surface Plasmon on Smooth and Rough Surfaces and on Gratings. Springer Tracts in Modern Physics, vol. 111. Springer, New York, NY (1988)
26. Zayats, A.V., Smolyaninov, I.I., Maradudin, A.A.: Nano-optics of surface plasmon polaritons. Phys. Rep. **408**(3–4), 131 (2005)

27. Thio, T., Pellerin, K.M., Linke, R.A., Lezec, H.J., Ebbesen, T.W.: Enhanced light transmission through a single subwavelength aperture. Opt. Lett. **26**(24), 1972 (2001)

28. Rivera, V.A.G., Ferri, F.A., Nunes, L.A.O., Zanatta, A.R., Marega Jr., E.: Focusing surface plasmons on Er^{3+} ions through gold planar plasmonic lenses. Appl. Phys. A **109**(4), 1037 (2012)

29. Carretero-Palacios, S., Mahboub, O., Garcia-Vidal, F.J., Martín-Moreno, L., Rodrigo, S.G., Genet, C., Ebbesen, T.W.: Mechanisms for extraordinary optical transmission through bull's eye structures. Opt. Express **19**(11), 10429 (2011)

30. Fu, Y., Zhou, X.: Plasmonic lenses: a review. Plasmonics **5**(3), 287 (2010)

31. Steele, J.M., Liu, Z., Wang, Y., Zhang, X.: Resonant and non-resonant generation and focusing of surface plasmons with circular gratings. Opt. Express **14**(12), 5664 (2006)

32. Coe, J.V., Heer, J.M., Teeters-Kennedy, S., Tian, H., Rodriguez, K.R.: Extraordinary transmission of metal films with arrays of subwavelength holes. Annu. Rev. Phys. Chem. **59**, 179 (2008)

33. Hsieh, B.-Y., Wang, N., Jarrahi, M.: Toward ultrafast pump-probe measurements at the nanoscale. Optic. Photon. News **22**(12), 48 (2011)

34. Srituravanich, W., Fang, N., Sun, C., Luo, Q., Zhang, X.: Plasmonic nanolithography. Nano Lett. **4**(6), 1085 (2004)

35. Shi, X., Hesselink, L.: Mechanisms for enhancing power throughput from planar nano-apertures of near-field optical data storage. Jpn. J. Appl. Phys. **41**, 1632 (2002)

36. Nahata, A., Linke, R.A., Ishi, T., Ohashi, K.: Enhanced nonlinear optical conversion from a periodically nanostructured metal flim. Opt. Lett. **28**(6), 423 (2003)

37. Liu, C., Kamaev, V., Vardeny, Z.V.: Efficiency enhancement of an organic light-emitting diode with a cathode forming two-dimensional periodic hole array. Appl. Phys. Lett. **86**, 143501 (2005)

38. Polman, A.: Plasmonics applied. Science **322**(5903), 868 (2008)

39. White, J.S., Veronis, G., Yu, Z., Barnard, E.S., Chandran, A., Fan, S., Brongersma, M.L.: Extraordinary optical absorption through subwavelength slits. Opt. Lett. **34**(5), 686 (2009)

40. Shalaev, V.: Engineering Space for Light with Metamaterials. NATO-ASI, Ottawa (2008)

41. Schröter, U., Heitmann, D.: Surface plasmon enhanced transmission through metallic gratings. Phys. Rev. B **58**, 15419 (1998)

42. Popov, E., Nevière, M., Enoch, S., Reinisch, R.: Theory of light transmission through subwavelength periodic hole arrays. Phys. Rev. B **62**, 16100 (2000)

43. Andrewartha, J.R., Fox, J.R., Wilson, I.J.: Further properties of lamellar grating resonance anomalies. Opt. Acta **26**(2), 197 (1979)

44. Barnes, W.L., Preist, T.W., Kitson, S.C., Sambles, J.R.: Physical origin of photonic energy gaps in the propagation of surface plasmons on gratings. Phys. Rev. B **54**, 6227 (1996)

45. Darmanyan, S.A., Zayats, A.V.: Light tunneling via resonant surface plasmon polariton states and the enhanced transmission of periodically nanostructured metal films: an analytical study. Phys. Rev. B **67**, 035424 (2003)

46. Wood, R.W.: Anomalous diffraction gratings. Phys. Rev. **48**, 928 (1935)

47. Chang, S.-H., Gray, S.K., Schatz, G.C.: Surface plasmon generation and light transmission by isolated nanoholes and arrays of nanoholes in thin metal films. Opt. Express **13**(8), 3150 (2005)

48. Bravo-Abad, J., Degiron, A., Przybilla, F., Genet, C., García-Vidal, F.J., Martín-Moreno, L., Ebbesen, T.W.: How light emerges from an illuminated array of subwavelength holes. Nat. Phys. **2**, 120 (2006)

49. Rivera, V.A.G., Ferri, F.A., Silva, O.B., Sobreira, F.W.A., Marega, E. Jr.: In: Kim K.Y. (ed.) Light transmission via subwavelength apertures in metallic thin films, Chapter 7, Intech, Croatia (2012)

50. Zia, R., Selker, M.D., Catrysse, P.B., Brongersma, M.L.: Geometries and materials for subwavelength surface plasmon modes. J. Opt. Soc. Am. A **21**(12), 2442 (2004)

51. Chang, D.E., Sørensen, A.S., Hemmer, P.R., Lukin, M.D.: Strong coupling of single emitters to surface plasmons. Phys. Rev. B **76**, 035420 (2007)
52. Ritchie, R.H.: Plasma losses by fast electrons in thin films. Phys. Rev. **106**, 874 (1957)
53. Oulton, R.F., Sorger, V.J., Genov, D.A., Pile, D.F.P., Zhang, X.: A hybrid plasmonic waveguide for subwavelength confinement and long-range propagation. Nat. Photon. **2**, 496 (2008)
54. Akimov, A.V., Mukherjee, A., Yu, C.L., Chang, D.E., Zibrov, A.S., Hemmer, P.R., Park, H., Lukin, M.D.: Generation of single optical plasmons in metallic nanowires coupled to quantum dots. Nature **450**, 402 (2007)
55. Rivera, V.A.G., Ledemi, Y., Osório, S.P.A., Ferri, F.A., Messaddeq, Y., Nunes, L.A.O., Marega Jr., E.: Optical gain medium for plasmonic devices. Proc. SPIE **8621**, 86211J (2013)
56. Gather, M.C., Meerholz, K., Danz, N., Leosson, K.: Net optical gain in a plasmonic waveguide embedded in a fluorescent polymer. Nat. Photon. **4**, 457 (2010)
57. García-Blanco, S.M., Sefunc, M.A., van Voorden, M.H., Pollnau, M.: Loss compensation in metal-loaded hybrid plasmonic waveguides using Yb^{3+} potassium double tungstate gain materials. doi: 10.1109/ICTON. 2012.6254494
58. Dahal, R., Ugolini, C., Lin, J.Y., Jiang, H.X., Zavada, J.M.: Erbium-doped GaN optical amplifiers operating at 1.54 μm. Appl. Phys. Lett. **95**(11), 111109 (2009)
59. Feigenbaum, E., Orenstein, M.: Modeling of complentary(void) plasmon waveguiding. J. Lightwave. Tech. **25**(9), 2547 (2007)
60. Veronis, G., Fan, S.: Modes of subwavelength plasmonic slot waveguides. J. Lightwave. Tech. **25**(9), 2511 (2007)
61. Yu, Z., Veronis, G., Fan, S., Brongersma, M.L.: Gain-induced switching in metal-dielectric-metal plasmonic waveguides. Appl. Phys. Lett. **92**, 041117 (2008)
62. García-Blanco, S.M., Pollnau, M., Bozhevolnyi, S.I.: Loss compensation in long-range dielectric-loaded surface plasmon-polariton waveguides. Opt. Express **19**(25), 25298 (2011)
63. Moreno, E., Rodrigo, S.G., Bozhevolnyi, S.I., Martín-Moreno, L., García-Vidal, F.J.: Guiding and focusing of electromagnetic fields with wedge plasmon polaritons. Phys. Rev. Lett. **100**, 023901 (2008)
64. Yan, M., Qiu, M.: Guided plasmon polariton at 2D metal corners. J. Opt. Soc. Am. B **24**(9), 2333 (2007)
65. Pile, D.F.P., et al.: Theoretical and experimental investigation of strongly localized plasmons on triangular metal wedges for subwavelength waveguiding. Appl. Phys. Lett. **87**, 061106 (2005)
66. Satuby, Y., Orenstein, M.: Surface-plasmon-polariton modes in deep metallic trenches-measurement and analysis. Opt. Exp. **15**(7), 4247 (2007)
67. Burke, J.J., Stegeman, G.I., Tamir, T.: Surface-polariton-like waves guided by thin, lossy metal films. Phys. Rev. B **33**, 5186 (1986)
68. Zia, R., Schuller, J.A., Brongersma, M.L.: Near-field characterization of guided polariton propagation and cutoff in surface plasmon waveguides. Phys. Rev. B **74**, 165415 (2006)
69. Verhagen, E., Polman, A., Kuipers, L.K.: Nanofocusing in laterally tapered plasmonic waveguides. Opt. Exp. **16**(1), 45 (2008)
70. Maier, S.A., et al.: Local detection of electromagnetic energy transport below the diffraction limit in metal nanoparticle plasmon waveguides. Nat. Mater. **2**, 229 (2003)
71. Onuki, T., et al.: Propagation of surface plasmon polariton in nanometer-sized metal-clad optical waveguides. J. Microsc. **210**(3), 284 (2003)
72. Conway, J.A., Sahni, S., Szkopek, T.: Plasmonic interconnects versus conventional interconnects: a comparison of latency, crosstalk and energy costs. Opt. Exp. **15**(8), 4474 (2007)
73. Brongersma, M. L., Kik, P. G.: Surface Plasmon Nanophotonics. Springer, P.O. Box 17, 3300 AA Dordrecht (2007)
74. Bharadwaj, P., Deutsch, B., Novotny, L.: Optical antennas. Adv. Opt. Photon. **1**(3), 438 (2009)

75. Schuck, P.J., Fromm, D.P., Sundaramurthy, A., Kino, G.S., Moerner, W.E.: Improving the mismatch between light and nanoscale objects with gold bowtie nanoantennas. Phys. Rev. Lett. **94**, 017402 (2005)

76. Søndergaard, T., Bozhevolnyi, S.I.: Metal nano-strip optical resonators. Opt. Exp. **15**(7), 4198 (2007)

77. Gersten, J., Nitzan, A.: Electromagnetic theory of enhanced Raman scattering by molecules adsorbed on rough surfaces. J. Chem. Phys. **73**(7), 3023 (1980)

78. Lee, B., Kim, S., Kim, H., Lim, Y.: The use of plasmonics in light beaming and focusing. Prog. Quantum. Electron. **34**(2), 47 (2010)

79. Editorial: Extending opportunities. Nat. Photon. **6**, 407 (2012)

80. Nielsen, M.A., Chuang, I.L.: Quantum Computation and Quantum Information. Cambridge Univ. Press, Cambridge (2000)

81. Gisin, N., Thew, R.: Quantum communication. Nat. Photon. **1**, 165 (2007)

82. Gramotnev, D.K., Bozhevolnyi, S.I.: Plasmonics beyond the diffraction limit. Nat. Photon. **4**, 83 (2010)

83. Zhang, J., Cai, L., Bai, W., Xu, Y., Song, G.: Hybrid plasmonic waveguide with gain medium for lossless propagation with nanoscale confinement. Opt. Lett. **36**(12), 2312 (2011)

84. Jin, X.R., Sun, L., Yang, X., Gao, J.: Quantum entanglement in plasmonic waveguides with near-zero mode indices. Opt. Lett. **38**(20), 4078 (2013)

85. Nezhad, M.P., Tetz, K., Fainman, Y.: Gain assisted propagation of surface plasmon polaritons on planar metallic waveguides. Opt. Exp. **12**(17), 4072 (2004)

86. Maier, S.A.: Gain-assisted propagation of electromagnetic energy in subwavelength surface plasmon polariton gap waveguides. Optic. Comm. **258**(2), 295 (2006)

87. Purcell, E.M.: Spontaneous emission probabilities at radio frequencies. Phys. Rev. **69**, 681 (1946)

88. Maier, S.A.: Effective mode volume of nanoscale plasmon cavities. Opt. Quant. Electron. **38** (1–3), 257 (2006)

89. Walls, D.F., Milburn, G.J.: Quantum Optics. Springer, New York, NY (2008)

90. Brune, M., et al.: Quantum Rabi oscillation: a direct test of field quantization in a cavity. Phys. Rev. Lett. **76**, 1800 (1996)

91. Finazzi, M., Ciaccacci, F.: Plasmon-photon interaction in metal nanoparticles: second-quantization perturbative approach. Phys. Rev. B **86**, 035428 (2012)

92. Archamabault, A., Marquier, F., Greffet, J.-J., Arnold, C.: Quantum theory of spontaneous and stimulated emission of surface plasmons. Phys. Rev. B **82**, 035411 (2010)

93. Trugler, A., Hohenester, U.: Strong coupling between a metallic nanoparticle and a single molecule. Phys. Rev. B **77**, 115403 (2008)

94. Van Vlack, C., Kristensen, P.T., Hughes, S.: Spontaneous emission spectra and quantum light-matter interactions from a strongly coupled quantum dot metal-nanoparticle system. Phys. Rev. B **85**, 075303 (2012)

95. Hummer, T., García-Vidal, F.J., Martín-Moreno, L., Zueco, D.: Weak and Strong coupling regimes in plasmonic QED. Phys. Rev. B **87**, 115419 (2013)

96. Monroe, C.: Quantum information processing with atoms and photons. Nature **416**, 238 (2002)

97. Naik, G.V., Kim, J., Boltasseva, A.: Oxides and nitrides as alternative plasmonic materials in the optical range. Opt. Mat. Exp. **1**(6), 1090 (2011)

98. Rivera, V.A.G., Ferri, F.A., Marega, E. Jr.: In: Kim KY (ed.) Localized Surface Plasmon Resonances: Noble Metal Nanoparticle Interaction with Rare-Earth Ions, (Chapter 11), Intech, Croatia (2012)

99. Dzsotjan, D., Käestel, J., Fleischhauer, M.: Dipole-dipole shift of quantum emitters coupled to surface plasmons of a nanowire. Phys. Rev. B **84**, 075419 (2011)

100. Rivera, V.A.G., Ledemi, Y., El-Amraoui, M., Messaddeq, Y., Marega Jr., E.: Resonant near-infrared emission of Er^{3+} ions in plasmonic arrays of subwavelength square holes. Proc. SPIE **8632**, 863225–1 (2013)

101. Rivera, V.A.G., Ledemi, Y., Messaddeq, Y., Marega Jr., E.: Plasmonic emission enhancement from Er^{3+}-doped tellurite glass via negative-nanobowtie. Proc. SPIE **8994**, 899421–1 (2014)
102. Silva, O.B., Rivera, V.A.G., Ferri, F.A., Ledemi, Y., Zanatta, A.R., Messaddeq, Y., Marega Jr., E.: Quantum-plasmonic interaction: emission enhancement of Er^{3+} - Tm^{3+} co-doped tellurite glass via tuning nanobowtie. Proc. SPIE **8809**, 88092X–1 (2013)
103. Kim, H., Bose, R., Shen, T.C., Solomon, G.S., Waks, E.: A quantum logic gate between a solid-state quantum bit and a photon. Nat. Photon. **7**, 373 (2013)
104. Lo, J.-W., Lien, W.-C., Lin, C.-A., He, J.-H.: Er-Doped ZnO nanorod arrays with enhanced 1540 nm emission by employing Ag island films and high-temperature annealing. ACS Appl. Mater. Interfaces **3**(4), 1009 (2011)
105. Mertens, H., Polman, A.: Plasmon-enhanced erbium luminescence. Appl. Phys. Lett. **89**(21), 211107 (2006)
106. Gopinath, A., Boriskina, S.V., Yerci, S., Li, R., Dal Negro, L.: Enhancement of the 1.54 μm Er^{3+} emission from quasiperiodic plasmonic arrays. Appl. Phys. Lett. **96**(7), 071113 (2010)
107. Hofmann, C.E., García de Abajo, F.J., Atwater, H.A.: Enhancing the radiative rate in III-V semiconductor plasmonic core-shell nanowire resonators. Nano Lett. **11**(2), 372 (2011)
108. Karaveli, S., Zia, R.: Spectral tuning by selective enhancement of electric and magnetic dipole emission. Phys. Rev. Lett. **106**, 193004 (2011)
109. Zhou, W., Dridi, M., Suh, J.Y., Kim, C.H., Co, D.T., Wasielewski, M.R., Schatz, G.C., Odom, T.W.: Lasing action in strongly coupled plasmonic nanocavity arrays. Nat. Nanotechnol. **8**, 506 (2013)
110. Zhou, W., Odom, T.W.: Tunable subradiant lattice plasmons by out-of-plane dipolar interactions. Nat. Nanotechnol. **6**, 423 (2011)
111. Gruner, T., Welsch, D.G.: Green-function approach to the radiation-field quantization for homogeneous and inhomogeneous Kramers-Kroning dielectrics. Phys. Rev. A **53**, 1818 (1996)
112. Sakurai, J.J.: Advanced Quantum Mechanics. Addison-Wesley, New York, NY (1967)
113. Judd, B.R.: Optical absorption intensities of rare earth ions. Phys. Rev. **127**(3), 750 (1962)
114. Gramotnev, D.K.: Adiabatic nanofocusing of plasmons by sharp metallic grooves: geometrical optics approach. J. Appl. Phys. **98**(10), 104302 (2005)
115. De Angelis, F., et al.: A hybrid plasmonic-photonic nanodevice for label-free detection of a few molecules. Nano Lett. **8**(8), 2321 (2008)
116. De Angelis, F., et al.: Nanoscale chemical mapping using three dimensional adiabatic compression of surface plasmon polaritons. Nat. Nanotechnol. **5**, 67 (2010)
117. Stockman, M.I.: Nanofocusing of optical energy in tapered plasmonic waveguides. Phys. Rev. Lett. **93**, 137404 (2004)
118. Bozhevolnyi, S.I., Volkov, V.S., Devaux, E., Laluet, J.-Y., Ebbesen, T.W.: Channel plasmon subwavelength waveguide components including interferometers and ring resonators. Nature **440**, 508 (2006)
119. Durach, M., Rusina, A., Stockman, M.I., Nelson, K.: Toward full spatiotemporal control on the nanoscale. Nano Lett. **7**(10), 3145 (2007)
120. Park, I.-Y., Kim, S., Choi, J., Lee, D.-H., Kim, Y.-J., Kling, M.F., Stockman, M.I., Kim, S.-W.: Plasmonic generation of ultrashort extreme ultraviolet light pulses. Nat. Photon. **5**, 677 (2011)
121. Issa, N.A., Guckenberger, R.: Optical nanofocusing on tapered metallic waveguides. Plasmonics **2**(1), 31 (2007)
122. Gramotnev, D.K., Pile, D.F.P., Vogel, M.W., Zhang, X.: Local electric field enhancement during nanofocusing of plasmons by tapered gap. Phys. Rev. B **75**, 035431 (2007)
123. Dereux, A., et al.: Direct interpretation of near-field optical images. J. Microsc. **202**(2), 320 (2001)
124. Mehtani, D., et al.: Optical properties and enhancement factors of the tips for apertureless near-field optics. J. Opt. A **8**(4), S183 (2006)

125. Chul Kim, H., Cheng, X.: Gap surface plasmon polaritons enhanced by plasmonic lens. Opt. Lett. **36**(16), 3082 (2011)
126. Rivera, V.A.G., Osório, S.P.A., Ledemi, Y., Manzani, D., Messaddeq, Y., Nunes, L.A.O., Marega Jr., E.: Localized surface plasmon resonance interaction with Er^{3+}-doped tellurite glass. Opt. Express. **18**(24), 25321 (2010)
127. Anger, P., Bharadwaj, P., Novotny, L.: Enhancement and quenching of single-molecule fluorescence. Phys. Rev. Lett. **96**, 113002 (2006)
128. Kühn, S., Håkanson, U., Rogobete, L., Sandoghdar, V.: Enhancement of single-molecule fluorescence using a gold nanoparticle as an optical nanoantenna. Phys. Rev. Lett. **97**, 017402 (2006)
129. Rivera, V.A.G., Ferri, F.A., Osório, S.P.A., Nunez, L.A.O., Zanatta Sr., A.R., Marega Jr., E.: Focusing surface plasmons on Er^{3+} ions with convex/concave plasmonic lenses. Proc. SPIE **8269**, 82692I (2012)
130. Muskens, O.L., et al.: Strong enhancement of radiative decay rate of emitters by single plasmonic nanoantennas. Nano Lett. **7**(9), 2871 (2007)
131. Halas, N.J., Lal, S., Chang, W.-S., Link, S., Nordlander, P.: Plasmons in strongly coupled metallic nanostructures. Chem. Rev. **111**(6), 3913 (2011)
132. Morton, S.M., Silverstein, D.W., Jensen, L.: Theoretical studies of plasmonics using electronic structure methods. Chem. Rev. **111**(6), 3962 (2011)
133. Moskovits, M., Jeong, D.H.: Engineering nanostructures for giant optical fields. Chem. Phys. Lett. **397**(1–3), 91 (2004)
134. Graeme McNay, David, E, Smith, W.E., Faulds, K., Graham, D.: Surface-enhanced Raman scattering (SERS) and surface-enhanced resonance Raman scattering (SERRS): a review of applications. Appl. Spec. **65**(8), 825 (2011)
135. Yariv, A., Yeh, P.: Optical Waves in Crystals: Propagation and Control of Laser Radiation. Wiley, Hoboken, NJ (2003)
136. Hao, J., et al.: Manipulating electromagnetic wave polarizations by anisotropic metamaterials. Phys. Rev. Lett. **99**, 063908 (2007)
137. Magruder, R.H., Wittig, J.E.: Wavelength tenability of the surface plasmon resonance of nanosize metal colloids in glass. J. Non-Cryst. Solids. **163**(2), 162 (1993)
138. Nolte, D.D.: Optical scattering and absorption by metal nanoclusters in GaAs. J. Appl. Phys. **76**(6), 3740 (1994)
139. Jana, N.R., Gearheart, L., Murphy, C.J.: Seeding growth for size control of 5-40 nm diameter gold nanoparticles. Langmuir **17**(22), 6782 (2001)
140. Sukharev, M., Seideman, T.: Phase and polarization control as a route to plasmonic nanodevices. Nano Lett. **6**(4), 715 (2006)
141. Eftekhari, F., Escobedo, C., Ferreira, J., Duan, X., Girotto, E.M., Brolo, A.G., Gordon, R., Sinton, D.: Nanoholes as nanochannels: flow through plasmonic sensing. Anal. Chem. **81** (11), 4308 (2009)
142. Curto, A. G. Ph.D. Thesis, Universitat Politècnica de Catalunya (2013)
143. Jackson, J.D. (ed.): Classical Electrodynamics. Wiley, Hoboken, NJ (1999)
144. Lindquist, N.C., et al.: Engineering metallic nanopartilces for plasmonics and nanophotonics. Rep. Prog. Phys. **75**(3), 036501 (2012)
145. Balanis, C.A. (ed.): Antenna Theory: Analysis and Design. Wiley, Hoboken, NJ (2005)
146. Chakrabarty, A., Wang, F., Minkowski, F., Sun, K., Wei, Q.H.: Cavity modes and their excitations in elliptical plasmonic patch nanoantennas. Opt. Exp. **20**(11), 11615 (2012)
147. Gutiérrez-Vega, J.C., Rodriguez-Dagnino, R.M., Meneses-Nava, M.A., Chávez-Cerda, S.: Mathieu functions, a visual approach. Am. J. Phys. **71**(3), 233 (2003)
148. Wang, F., Chakrabarty, A., Minkowski, F., Sun, K., Wei, Q.-H.: Polarization conversion with elliptical patch nanoantennas. Appl. Phys. **101**(2), 023101 (2012)
149. Gong, X., Chen, Y., Lin, Y., Tan, Q., Luo, Z., Huang, Y.: Polarized spectral analysis and laser demonstration of $Nd^{3+}:Bi^2 (MoO_4)_3$ biaxial crystal. J. Appl. Phys. **103** (2008).

150. Bergman, D.J., Stockman, M.I.: Surface plasmon amplification by stimulated emission radiation: quantum generation of coherent surface plasmons in nanosystems. Phys. Rev. Lett. **90**, 027402 (2003)
151. Stockman, M.I.: Spasers explained. Nat. Photon. **2**, 327 (2008)
152. Noginov, M.A., et al.: Demonstration of a spaser-based nanolaser. Nature **460**, 1110 (2009)

Chapter 4
Potential Applications

The focus of this book was on introducing solid concepts about collective plasmon modes in gain media, centered on the interaction/coupling between a quantum emitter (QE) (created by rare-earth ions (REI)) with a metallic nanostructure. This quantum–plasmonic interaction results in a polarization of the metallic electrons, or, in the case of transition between electronic levels of the emitter, a sustained field enhancement resulting from resonance coupling or an energy transfer owing to nonresonance coupling by the metallic nanostructures and the optically active surrounding medium. In other words, we have the ability to manipulate the electronic states and decay pathways available in a QE at the nanoscale, thus increasing or decreasing the radiative rate of a QE. In addition, by overcoming the SPP propagation losses within a gain medium, the propagation can be sustained owing to efficient coupling between a surface plasmon polariton (SPP) and the QE. Alternatively, the simplest solution is to replace the dielectric substrate for another dielectric that possesses a higher refractive index.

From this point of view, we have the ability to engineer a way to counteract the absorption losses in metal, to enhance luminescence, and to control the polarization and phase of the QEs. This provides us with exciting applications of SPPs and localized surface plasmon resonance (LSPR) in nanophotonics, with tremendous commercialization potential in the following areas. (1) Optical telecommunication, owing to their capacity for integrated optics at the nanoscale. (2) Sensing, since the spectral position of the resonances of a metallic nanostructure depends on the dielectric environment within the EM near field and can be improved by the presence of an REI and its unique frequency-converting capability (both down-conversion and up-conversion). (3) Clean energy, obtained by increasing the response of a p-n junction, by means of localized SPPs or LSPR in REI-doped materials. In Fig. 4.1, we present a roadmap highlighting some exciting and as-yet-unexplored topics related to these points, with the promise of achieving the next stage of research in quantum plasmonics, using REI-doped materials (a gain medium generating hybrid SPP/LSPR and loss compensation) with a very fruitful and productive future.

© V.A.G. Rivera, O.B. Silva, Y. Ledemi, Y. Messaddeq, and E. Marega Jr. 2015   117
V.A.G. Rivera et al., *Collective Plasmon-Modes in Gain Media*, SpringerBriefs
in Physics, DOI 10.1007/978-3-319-09525-7_4

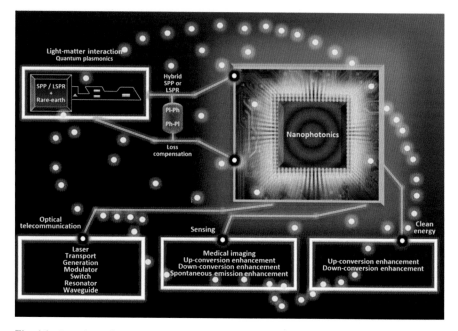

Fig. 4.1 A roadmap for quantum plasmonics based on the interactions of SPP/LSPR with REIs. Three *big areas* are highlighted owing to their technological importance and benefits to society

In fact, these applications are indeed exciting motivation, despite the fact that we do not know exactly what form future quantum–plasmonic technologies will take. However, we can say that in the photonic quantum technologies being developed, it is most likely that quantum states of light will be transmitted and that information processing will be performed on these states. It is worth saying that if we are to realize these technologies, then we will need to constantly exploit the latest developments in the nanophotonics field.

Currently, we have a variety of technological routes open in the photonics field, since there are a great variety of QE candidates to be employed in the development of new light sources for nanodevices, such as lasers, LEDs, and single-photon sources, but also in chemistry and life sciences where they act as nanoscopic probes and labels.

Nonetheless, REIs are, from our point view, the most promising because of their unique characteristics, such as their magnetic, luminescent, and electrochemical properties. They suit newly emerging devices for their greater efficiency, reduced weight, high speed, greater durability, and their miniaturization potential. In this route, nanophotonics is not only about very small photonic circuits and nanochips, but also about new ways of sculpting the flow of light by means of nanostructures and/or plasmonic NPs exhibiting fascinating optical properties that we are learning to manipulate and control. Therefore, a deep investigation of these effects may provide us with new perspective on plasmonics and may yield many more future applications, as listed in Fig. 4.1. This can be very helpful for developing applications of such metallic nanostructures.

The dynamic components reviewed in this book will demonstrate the first realizations of the unique features inherent to quantum–plasmonic interactions in a gain medium, with benefits for the technological fields of optical telecommunications, sensing, and clean energy. This quantum information can also provide us with pathways for guiding photon across a chip in whatever direction the designers desire.

Ross Stanley has shown that plasmonic nanostructures can be employed to support the field of mid-infrared (MIR) photonics [1]. In the beginning, metallic nanostructures were studied in the visible region, but they are now being exploited to enhance the performance of MIR sources, sensors, and detectors for applications such as thermal imaging and chemical sensing. It is expected that a better understanding of the geometric and dielectric impedance will open new windows to their successful development. In Stanley's view, the challenge now is to integrate near/mid-infrared plasmonics technologies in cost-effective, compact, and consistent platforms.

Today's microprocessors use ultrafast transistors with dimensions on the order of 22 nm. Although it is now routine practice to produce fast and small transistors, there is a major problem with carrying digital information to the other end of a microprocessor (a few centimeters away). Plasmonic circuits can merge electronics and photonics at the nanoscale, and offers us a solution to this size-compatibility problem. Using plasmonic chips (with light travelling on a nanowire) opens up a path toward quantum control of nanophotonics via the generation, manipulation, and detection of quantum information (remembering that the photon is the fundamental unit of information).

Excellent reviews on plasmonic circuitry can be found in Refs. [2–4] and in the famous paper by Harry A. Atwater about the promise of plasmonics [5]. In his words, "a technology that squeezes EM waves into minuscule structures may yield a new generation of superfast computer chips and ultrasensitive molecular detectors."

Ultimately, practical photonic circuits use a combination of plasmonic and dielectric components, taking advantage of the best performance available. In other words, whether or not the outcome will lead to hybrid SPP/LSPR circuits, the activity stimulated by plasmonics and REI research will bring the long-held dream of an integrated optical circuit ever closer to reality, as in Fig. 4.1. Of course, there are countless routes to merging the best features of photonics and plasmonics, with the only limits being our imagination and our determination to realize the promise of nanophotonic technology.

References

1. Stanley, R.: Plasmonics in the mid-infrared. Nat. Photon. **6**, 409 (2012)

2. Ozbay, E.: Plasmonics: Merging Photonics and Electronics at Nanoscale Dimensions. Science **311**, 189 (2006)
3. Ebbesen, T.W., Genet, C., Bozhevolnyi, S.I.: Surface-plasmon circuitry. Phys. Today 44–50 (2008)
4. Tame, M.S., McEnery, K.R., Ozdemir, S.K., Lee, J., Maier, S.A., Kim, M.S.: Quantum plasmonics. Nat. Phys. **9**, 329 (2013)
5. Atwater, H.A.: The promise of plasmonics. Scientific American INC. 56–63 (2007)

Index

© V.A.G. Rivera, O.B. Silva, Y. Ledemi, Y. Messaddeq, and E. Marega Jr. 2015   121
V.A.G. Rivera et al., *Collective Plasmon-Modes in Gain Media*, SpringerBriefs
in Physics, DOI 10.1007/978-3-319-09525-7